Introduction to Arnold's Proof of the Kolmogorov–Arnold–Moser Theorem

This book provides an accessible step-by-step account of Arnold's classical proof of the Kolmogorov–Arnold–Moser (KAM) Theorem. It begins with a general background of the theorem, proves the famous Liouville–Arnold theorem for integrable systems and introduces Kneser's tori in four-dimensional phase space. It then introduces and discusses the ideas and techniques used in Arnold's proof, before the second half of the book walks the reader through a detailed account of Arnold's proof with all the required steps. It is a useful guide for advanced students of mathematical physics, in addition to researchers and professionals.

Features

- Applies concepts and theorems from real and complex analysis (e.g., Fourier series and implicit function theorem) and topology in the framework of this key theorem from mathematical physics

- Covers all aspects of Arnold's proof, including those often left out in more general or simplified presentations

- Discusses, in detail, the ideas used in the proof of the KAM theorem and puts them in historical context (e.g., mapping degree from algebraic topology).

Achim Feldmeier is a professor at Universität Potsdam, Germany.

Introduction to Arnold's Proof of the Kolmogorov–Arnold–Moser Theorem

Achim Feldmeier

CRC Press
Taylor & Francis Group
Boca Raton London New York

CRC Press is an imprint of the
Taylor & Francis Group, an **informa** business

First edition published 2023
by CRC Press
6000 Broken Sound Parkway NW, Suite 300, Boca Raton, FL 33487-2742

and by CRC Press
4 Park Square, Milton Park, Abingdon, Oxon, OX14 4RN

CRC Press is an imprint of Taylor & Francis Group, LLC

© 2022 Achim Feldmeier

Library of Congress Cataloging-in-Publication Data

ISBN: 978-1-032-26065-5 (hbk)
ISBN: 978-1-032-26338-0 (pbk)
ISBN: 978-1-003-28780-3 (ebk)

DOI: 10.1201/9781003287803

Typeset in NimbusRomaNo9L-Regu font
by KnowledgeWorks Global Ltd.

Publisher's note: This book has been prepared from camera-ready copy provided by the authors.

To my parents

Contents

Preface ix

Chapter 1   Hamilton Theory 1

      1.1   Hamilton Equations 1
      1.2   Generating Functions 2
      1.3   Infinitesimal Canonical Transformation 4
      1.4   Conserved Phase-Space Measure 6
      1.5   Coordinate Relations 8
      1.6   Action-Angle Variables 9
      1.7   Integrable Systems and Phase-Space Tori 11
      1.8   Kneser's Theorems for Four-Dimensional Phase Space 31
      1.9   Celestial Mechanics 34

Chapter 2   Preliminaries 39

      2.1   Notation 39
      2.2   Domain Reduction 43

Chapter 3   Outline of the KAM Proof 47

      3.1   Perturbation Series vs. Fast Iteration 48
      3.2   Perturbation Theory with New Frequencies 55
      3.3   Frequency Diffeomorphism 58
      3.4   Diophantine Condition 60
      3.5   Fast Iteration Plus Diophantine Condition 62
      3.6   Resonance Cutoff 64
      3.7   Domains without Interior 67
      3.8   Analytic Functions 70
      3.9   Motion on a Torus 71
      3.10 Small Numbers 72

Chapter 4   Proof of the KAM Theorem 75

      4.1   Fundamental Theorem 76
      4.2   Inductive Theorem 89
      4.3   KAM Theorem 101

Chapter 5   Analytic Lemmas 115

      5.1   Lemma A1. Exponentials vs. Powers 115
      5.2   Functions in Several Complex Variables 116

5.3   Lemma A2. Fourier Coefficients 118
5.4   Lemma A3. Cauchy Estimates 122
5.5   Lemma A4. Lagrange and Taylor Estimates 124

Chapter 6   Geometric Lemmas 127

6.1   An Implicit Function Theorem 127
6.2   Mapping Degree 129
6.3   Lemma G1. Surjectivity of Maps Near Identity 131
6.4   Brouwer's Lemma 136
6.5   Modern Proof of Lemma G1 143
6.6   Lemma G2. Frequency Diffeomorphism 145
6.7   Lemma G3. Canonical Transformation 148
6.8   Lemma G4. Domain Size of a Bounded Map 153
6.9   Lemma G5. Frequency Variation Lemma 154

Chapter 7   Convergence Lemmas 161

7.1   Lemma C1. Iterated Canonical Transformation 161
7.2   Lemma C2. Integral Curves and Segments 163
7.3   The Push Forward 164
7.4   Lemma C3. Limiting Flow 169
7.5   Lemma C4. Measure of the Limit 171

Chapter 8   Arithmetic Lemmas 175

8.1   Lemma D1. Measure of Boundary Layer 175
8.2   Lemma D2. Measure of Inner Layer 176
8.3   Lemma D3. Measure of Strip 177
8.4   Lemma D4. Measure of Layers and Strips 178
8.5   Lemma D5. Integral Vectors with Given Norm 180
8.6   Lemma D6. Integral Vectors below Given Norm 182
8.7   Lemma D7. Measure of Lost Domain 182

References 185

Person Index 193

Subject Index 197

Preface

This book gives a step-by-step presentation of Arnold's proof of the KAM theorem from 1963. The theorem was first announced by Kolmogorov in 1954 and was independently proved, in a somewhat different form, by Moser (1962). KAM is an acronym for Kolmogorov, Arnold and Moser.

The theorem is among the main achievements of mathematical physics in the 20th century. It states (at first sight, somewhat tautologically) that integrable Hamiltonian systems subject to small perturbations are nearly-integrable systems:

> Nearly-integrable Hamiltonian systems [are] Hamiltonian systems which are small perturbations of an integrable system and which, in general, exhibit a much richer dynamics than the integrable limit. Nevertheless, KAM theory asserts that, under suitable assumptions, the majority (in the measurable sense) of the initial data of a nearly-integrable system behaves as in the integrable limit. (Chierchia 2009, p. 5064)

Each phase-space trajectory of an integrable Hamiltonian system[1] lies on a torus of dimension $m \leq n$ in $2n$-dimensional phase space, where $n \in \mathbb{N}$ is the number of degrees of freedom; we assume torus dimension $m = n$ here (for systems in involution). The motion on the torus is then quasi-periodic, $q_i(t) = \omega_i t + q_i(0)$, with n angle variables q_i, n angular frequencies ω_i and time t. The tori fill phase space densely, and each torus is characterized by n constants of motion (e.g., energy). The following two quotations give short descriptions of the KAM theorem.

> For a small perturbation [...], most of the invariant tori with incommensurable frequencies [...] do not disappear, but are merely slightly deformed. The trajectories of the perturbed motion beginning on the deformed torus [...] fill it everywhere densely and conditionally periodically. (Arnold 1963b, p. 100)

> The 'classical' KAM Theorem establishes persistence of invariant Lagrang[i]an tori in nearly integrable Hamiltonian systems. These tori are quasi-periodic with Diophantine frequency vectors and their union is a nowhere dense set of positive measure in the phase space. (Broer & Sevryuk 2010, p. 280)

---

[1] A Hamiltonian system is fully specified by its Hamilton function $H(p,q)$, whether it is a planetary system or an ensemble of molecules. The Hamilton function is supposed to be an analytic function periodic in the position variables q.

The trajectories of the slightly perturbed systems remain restricted to a low-dimensional torus, and the motion remains deterministic. By contrast, in a thermo-dynamic system the trajectories explore large portions of phase space in a random, stochastic manner. For example, if the ergodic hypothesis holds and energy is the only conserved quantity, trajectories should lie densely in a $(2n-1)$-dimensional subspace of $2n$-dimensional phase space.

The KAM theorem therefore shows that periodic, deterministic motion, as studied in classical mechanics including celestial mechanics, is stable under sufficiently small perturbations 'for all times,' and thus does not give over to stochastic, thermal behavior under slightest perturbations, as was the dominant view in the early 20th century.

[U]ntil the fifties it was a common belief that arbitrarily small perturbations can turn an integrable system into an ergodic one on each energy surface. In the twenties there even appeared an – erroneous – proof of this 'ergodic hypothesis' by Fermi. But in 1954 Kolmogorov observed that the converse is true – the majority of tori survives. (Pöschel 2001, p. 4)

[T]herefore such systems are still not ergodic contrary to the hope, widespread until recently, that this would not be the case. (Gallavotti 1983, p. 360)

Arnold's statement and proof of the KAM theorem cover roughly 15 pages in (1963a), and a little more in (1963b), which is largely the same proof. To my knowledge, the full proof is so far not covered in a textbook, but parts of it can be found, e.g., in Arnold & Avez (1968), Thirring (1977), Rüssmann (1979) and Chierchia (2009).[2]

This is somewhat surprising since Arnold's proof uses almost only standard methods from real and complex analysis, as taught in the first two years of university courses in mathematics. Yet, these methods are applied in a quite intricate way. To give an example, the convergence of domains (to sets of finite measure) and functions (to analytic functions) in an infinite sequence of canonical transformations is obtained by introducing seven mutually dependent sequences of 'control' numbers, $M, \delta, \beta, \gamma, \rho, \theta, \Theta$, which progress to zero geometrically.

In his book, *The KAM Story*, H. S. Dumas observes:

Despite the elementary nature of individual steps in the proofs, and despite occasional statements that the proofs are 'trivial,' in fact the global details of KAM proofs can be formidable [...] This is especially true for otherwise mathematically literate scientists [...] This last group could include a significant

---

[2]Moser's proof is treated in, e.g., Sternberg (1969), Siegel & Moser (1971) and Arrowsmith & Place (1990).

fraction of research physicists, even some theoretical physicists [...] Elementary rigorous analysis is certainly not beyond their ability, but many will simply never have spent the time required to master the techniques, and until they do, the 'hard analysis' in KAM proofs could seem all but impenetrable. I think a lot of the reputed difficulty of KAM proofs stems from this unfortunate fact. (Dumas 2014, pp. 109-110)

M. Sevryuk, in Khesin & Tabachnikov (2014), p. 181, states that

[...] these studies culminated in Arnold's famous (and technically extremely hard) result.

And John Hubbard (2007), p. 215, writes in the Kolmogorov heritage volume:

During these [earlier] years, I tackled all the proofs [of the KAM theorem] that I knew: Arnold's, Moser's, Sternberg's [...] I did not succeed in mastering a single one. And I am far from being alone: I know numerous dynamicists who realize that they ought be able to prove the theorem, who even teach it sometimes, but who have never mastered the proof either. [References omitted]

There has to be some justification for studying in extenso – slightly more than half of the present book is devoted to it – a 60-year-old proof. My reasons for doing so are the following: (i) I believe that Arnold's proof is one of the classics of 20th century mathematics, on a par with Gödel's incompleteness theorem and Turing's theory of the universal register machine. (ii) I consider Arnold's proof as (still) the best entry point to KAM theory. All that is required is calculus of several complex variables (plus a little degree theory from algebraic topology), as taught in the first three or four semesters of a bachelor study in physics or mathematics. (iii) Arnold's proof is exceptional in that it covers the full physical situation of a mechanical (Hamiltonian) system. Moser's proof deals with an idealized mathematical situation; and many of the modern presentations of the KAM theorem leave out the severe complications introduced by making the frequency map $A : p \mapsto \omega$ a diffeomorphism at each iteration step (on a shrinking domain). (iv) Arnold's proof itself is quite brief and terse, leaving out many details and minor steps.

Being a theoretical physicist myself, I have written mainly for physicists. Therefore, deep and subtle mathematical concepts like, e.g., the mapping degree are only introduced using elementary arguments and historical proofs, without going into technical details.

I thank the two anonymous reviewers of the book for their positive feedback. I thank Professor Arkadi Pikovski for helpful discussions and Rebecca Hodges-Davies at Taylor & Francis for her friendly support. Most of all, I thank my wife, Gudrun.

For feedback, please contact me by email: afeld@uni-potsdam.de.

Berlin, March 2022

1 Hamilton Theory

1.1 HAMILTON EQUATIONS

The principle of least action is

$$\delta \int_{t_1}^{t_2} dt\, L(q,\dot{q}) = 0, \tag{1.1}$$

with time t, generalized positions $q_1, \ldots, q_n$ for a system with n degrees of freedom, generalized speeds $\dot{q}_1, \ldots, \dot{q}_n$, Lagrange function L (not further specified here, but assumed to not depend explicitly on t) and variational derivative δ (a total derivative, where $\delta q(t)$ is an arbitrary, infinitesimal function).

The q_i and $\dot{q}_i$ appear as independent variables in L, and are thought of as Cartesian coordinates in a $2n$-dimensional velocity phase space (the dot in $\dot{q}_i$ is then merely a part of the variable name). Equation (1.1) fixes the functions $q_i(t)$ and $\dot{q}_i(t) = dq_i/dt$; its solution $(q_i(t), \dot{q}_i(t))$ is a curve with parameter t in phase space, the system trajectory.

Alternatively, the $q_i(t)$ and $\dot{q}_i(t)$ can be treated as independent functions (the following approach is due to Hilbert, see Kneser 1921, p. 279, and Nordheim & Fues 1927, p. 92), where the $\dot{q}_i$ are subject to n constraints

$$\dot{q}_k - r_k = 0. \tag{1.2}$$

Multiplying these constraints with Lagrange multipliers λ_i and adding the resulting terms to Eq. (1.1) gives

$$\delta \int dt \left[L(q,r) + \sum_{i=1}^{n} \lambda_i(\dot{q}_i - r_i) \right] = 0, \tag{1.3}$$

Assuming independent variations δq_i and δr_i, one has

$$\int dt \sum_{i=1}^{n} \left[\frac{\partial L}{\partial q_i} \delta q_i + \left(\frac{\partial L}{\partial r_i} - \lambda_i \right) \delta r_i \right] = 0. \tag{1.4}$$

By independence of the δq_i and δr_i, the round bracket must vanish, or

$$\lambda_i = \frac{\partial L}{\partial r_i}. \tag{1.5}$$

(The somewhat odd equations $\partial L/\partial q_i = 0$ are also correct, if all $\delta \dot{q}_i = d(\delta q_i)/dt$ are resolved into δq_i.) Putting this in Eq. (1.3) gives

$$\delta \int dt \left[L(q,r) + \sum_{i=1}^{n} \frac{\partial L}{\partial r_i}(\dot{q}_i - r_i) \right] = 0. \tag{1.6}$$

DOI: 10.1201/9781003287803-1

1

Introducing

$$p_i = \frac{\partial L}{\partial r_i} \quad \text{and} \quad H(p,q) = \sum_{i=1}^{n} p_i r_i - L(q,r), \tag{1.7}$$

one obtains

$$\delta \int dt \left[\sum_{i=1}^{n} p_i \dot{q}_i - H(p,q) \right] = 0, \tag{1.8}$$

which is the form of the principle of least action introduced by Hamilton. Assuming interchangeability of δ and d/dt and performing an integration by parts assuming vanishing variations at t_1 and t_2, one finds (with $H_{p_i} = \partial H / \partial p_i$, etc.)

$$0 = \int dt \sum_{i=1}^{n} [\delta p_i \, \dot{q}_i + p_i \, \delta \dot{q}_i - H_{p_i} \, \delta p_i - H_{q_i} \, \delta q_i]$$

$$= \int dt \sum_{i=1}^{n} [\delta p_i \, \dot{q}_i - \dot{p}_i \, \delta q_i - H_{p_i} \, \delta p_i - H_{q_i} \, \delta q_i] . \tag{1.9}$$

Assuming independent δp_i and δq_i, one obtains the Hamilton equations,

$$\dot{p}_i = -\frac{\partial H}{\partial q_i}, \qquad\qquad \dot{q}_i = \frac{\partial H}{\partial p_i} \tag{1.10}$$

with $i = 1, \ldots, n$. The p_i, q_i are called *canonical variables* or canonical coordinates. The first set of equations corresponds to Newton's second law, the second set to the Legendre transformation from variables $\dot{q}_i$ to p_i.

Throughout this book, phase space will be $\mathbb{R}^{2n}$ equipped with Cartesian coordinates $p_1, \ldots, p_n, q_1 \ldots, q_n$.[1] For given initial conditions $p_i(0), q_i(0)$, there is a unique solution curve $p_i(t), q_i(t)$ for $t \geq 0$ of the first-order system (1.10) in phase space.

1.2   GENERATING FUNCTIONS

We include now time t as an explicit variable in H and L. This is only required on a more practical side, to obtain a new Hamilton function K (see Eq. 1.20). If this is not intended, one needs to consider only $p_i \, dq_i$ in the following instead of $p_i \, dq_i - H \, dt$ (see Nordheim & Fues 1927, p. 100).

A *canonical transformation* is a change of canonical variables p_i, q_i of a Hamiltonian $H(p,q,t)$ to new variables P_i, Q_i that are again canonical, i.e., obey the Hamilton equations with a new Hamilton function $K(P,Q,t)$.

The Hamilton equations are obtained from the principle of least action,

$$\delta W = \delta \int_{t_1}^{t_2} dt \left[\sum_{i=1}^{n} p_i \dot{q}_i - H(p,q,t) \right] = 0. \tag{1.11}$$

---

[1] In Hamiltonian mechanics, the momenta p_i are usually listed before the positions q_i, due to the central importance of the 1-form $\sum_i p_i \, dq_i$ and the 2-form $\sum_i dp_i \wedge dq_i = d(\sum_i p_i \, dq_i)$. Only a few of the following formulas will be coordinate-free.

Therefore, the new variables P_i, Q_i are canonical if also

$$\delta V = \delta \int_{t_1}^{t_2} dt \left[\sum_{i=1}^{n} P_i \dot{Q}_i - K(P,Q,t) \right] = 0. \tag{1.12}$$

Assume that the two action integrals (without variation) are related by

$$W - V = F(t_2) - F(t_1), \tag{1.13}$$

where $F(P,Q,p,q,t)$ is an arbitrary function, in which however only $2n$ of the $4n$ canonical variables are independent (see below). Then, since all variations at t_1 and t_2 vanish by assumption, one has

$$\delta W - \delta V = \delta F(t_2) - \delta F(t_1) = 0. \tag{1.14}$$

Then $\delta W = 0$ implies δV, i.e., if p,q are canonical variables, so are P,Q. Write Eq. (1.13) as

$$W - V = \int_{t_1}^{t_2} dt \, \frac{dF}{dt} \tag{1.15}$$

and assume that

$$F = F(Q,q,t), \tag{1.16}$$

with dependent variables $P = P(q,Q)$ and $p = p(q,Q)$. Then Eq. (1.13) gives

$$\int_{t_1}^{t_2} dt \left[\sum_i p_i \dot{q}_i - \sum_i P_i \dot{Q}_i - H + K \right] = \int_{t_1}^{t_2} dt \left[\sum_i F_{q_i} \dot{q}_i + \sum_i F_{Q_i} \dot{Q}_i + F_t \right]. \tag{1.17}$$

Since F is an arbitrary function, we can equate the integrands without loss of generality,

$$\sum_i p_i \dot{q}_i - \sum_i P_i \dot{Q}_i - H + K = \sum_i F_{q_i} \dot{q}_i + \sum_i F_{Q_i} \dot{Q}_i + F_t. \tag{1.18}$$

From this follows, by independence of the canonical variables,

$$p_i = \frac{\partial F}{\partial q_i}, \qquad\qquad P_i = -\frac{\partial F}{\partial Q_i} \tag{1.19}$$

with $i = 1,\ldots,n$; furthermore,

$$K(P,Q,t) = H(p,q,t) + F_t(q,Q,t). \tag{1.20}$$

Equation (1.19) is the simplest possible canonical transformation. In most texts, including Arnold (1963a,b), a slightly different transformation is used. To obtain this, use again Eq. (1.15), but assume now that

$$F = -\sum_i P_i Q_i + S(P,q,t) \tag{1.21}$$

with arbitrary S. Then

$$\int_{t_1}^{t_2} dt \left[\sum_i p_i \dot{q}_i - \sum_i P_i \dot{Q}_i - H + K \right]$$
$$= \int_{t_1}^{t_2} dt \left[-\sum_i P_i \dot{Q}_i - \sum_i \dot{P}_i Q_i + \sum_i S_{P_i} \dot{P}_i + \sum_i S_{q_i} \dot{q}_i + S_t \right], \qquad (1.22)$$

which is fulfilled if

$$p_i = \frac{\partial S}{\partial q_i}, \qquad\qquad Q_i = \frac{\partial S}{\partial P_i} \qquad (1.23)$$

for $i = 1, \ldots, n$ (the formula $p_i = \partial S / \partial q_i$ occurs, e.g., in Jacobi 1884, lecture 20), and

$$K(P, Q, t) = H(p, q, t) + S_t(P, q, t). \qquad (1.24)$$

Arnold (1963a) uses S in an infinitesimal canonical transformation 'near the identity,' defined by

$$S(P, q) = \sum_i P_i q_i + s(P, q) \qquad (1.25)$$

with $\|s\| \ll \|S\|$. Then $(i = 1, \ldots, n)$

$$p_i = P_i + \frac{\partial s}{\partial q_i}, \qquad\qquad Q_i = q_i + \frac{\partial s}{\partial P_i}. \qquad (1.26)$$

Arnold assumes an infinite sequence (S_s) of infinitesimal canonical transformations and proves (Arnold 1963a, sect. 4.4.1°) uniform convergence $S_s \to S_\infty$.

We summarize Eq. (1.15) and Eq. (1.18) in a lemma.

Lemma 1 *If p_i, q_i are canonical variables with Hamilton function H and*

$$\left[\sum_i p_i \, dq_i - H(p, q, t) \, dt \right] - \left[\sum_i P_i \, dQ_i - K(P, Q, t) \, dt \right] = dF, \qquad (1.27)$$

i.e., the difference of the two 1-forms in brackets is the total differential of an arbitrary function F, then P_i, Q_i are canonical variables with Hamilton function K.

Specifically, F is to be understood as $F = F(Q, q, t)$. If F depends on other combinations of new and old canonical variables, as in Eq. (1.21), one can rewrite this as a function $F(Q, q, t)$ by using the (inverse) relations $p(P, Q)$ and $q(P, Q)$.

1.3 INFINITESIMAL CANONICAL TRANSFORMATION

Consider the infinitesimal canonical transformation $(i = 1, \ldots, n)$

$$p_i \to P_i = p_i + \varepsilon \pi_i(p, q),$$
$$q_i \to Q_i = q_i + \varepsilon \eta_i(p, q) \qquad (1.28)$$

for small ε. The functions π and η shall not depend on time, thus the old and new Hamilton function are identical, and

$$\sum_i P_i \, dQ_i - \sum_i p_i \, dq_i = \varepsilon \, dF. \tag{1.29}$$

Here ε was introduced to make the right side small of second order: for an arbitrary canonical transformation, the left side is small of first order, but for Eq. (1.28), it is small of second order (see below). Inserting Eq. (1.28) in (1.29) gives

$$\sum_i (p_i + \varepsilon \pi_i)(dq_i + \varepsilon \, d\eta_i) - \sum_i p_i \, dq_i = \varepsilon \, dF \tag{1.30}$$

or, to first order in ε, by independence of canonical variables,

$$\pi_i \, dq_i + p_i \, d\eta_i = dF \tag{1.31}$$

for $i = 1, \ldots, n$. Therefore,

$$\sum_i \pi_i \, dq_i - \sum_i \eta_i \, dp_i = dF - d\sum_i p_i \, \eta_i = -dK, \tag{1.32}$$

where $K = K(p,q)$. Inserting Eq. (1.32) in (1.28) gives $(i = 1, \ldots, n)$

$$\frac{P_i - p_i}{\varepsilon} = -\frac{\partial K}{\partial q_i}, \qquad \frac{Q_i - q_i}{\varepsilon} = \frac{\partial K}{\partial p_i}. \tag{1.33}$$

Identifying $H \equiv K$, $dt \equiv \varepsilon$, $dp_i \equiv P_i - p_i$, $dq_i \equiv Q_i - q_i$, this allows to interpret temporal evolution of a mechanical system as a sequence of infinitesimal canonical transformations. (This also allows to deal with p, q alone in Hamiltonian mechanics, treating time as a mere parameter of trajectories.)

With the definition of the *Poisson* bracket of two functions $f(p,q)$ and $g(p,q)$ of the canonical variables,

$$\{f,g\} = \sum_{i=1}^n \left(\frac{\partial f}{\partial p_i} \frac{\partial f}{\partial q_i} - \frac{\partial f}{\partial q_i} \frac{\partial f}{\partial p_i} \right), \tag{1.34}$$

the change in f by the infinitesimal canonical transformation generated by K from Eq. (1.33) is

$$df = \sum_i \frac{\partial f}{\partial p_i} \, dp_i + \sum_i \frac{\partial f}{\partial q_i} \, dq_i = -\varepsilon \sum_i \frac{\partial f}{\partial p_i} \frac{\partial K}{\partial q_i} + \varepsilon \sum_i \frac{\partial f}{\partial q_i} \frac{\partial K}{\partial p_i},$$

or

$$df = \varepsilon \, \{K, f\}. \tag{1.35}$$

This can be read as

$$\frac{df}{dt} = \{K, f\}. \tag{1.36}$$

The differential df in the context of a canonical transformation requires some clarification. Like any transformation, a canonical transformation $(p,q) \rightarrow (P,Q)$ allows to express one and the same function at one and the same point in phase space in two different coordinate systems or variables, $f(p,q) = f(p(P,Q), q(P,Q)) = F(P,Q)$.[2] By contrast, $df = f' \, dt$ is the change of f under a tiny (numerical) change of its argument t. The relation of these two aspects (as it occurs in Eq. 1.35) is discussed in Goldstein et al. (2002), p. 401:

> If we pose the question, How does a function change under a canonical transformation? the answer depends on whether we should take an active or a passive point of view. From the passive point of view, the function changes in form, or in functional dependence, but it does not change in value. This is because in general the function, call it U, has a different functional dependence on (Q,P) than it does on (q,p). Its value however remains the same [... since] both sets of coordinates refer to the same physical location in phase space but use different coordinates to describe the phase space.
>
> In contrast to this, if we consider the canonical transformation from an active point of view, then we are talking about a translation of the system from point A to point B, from position (q_A, p_A) to position (q_B, p_B). From this point of view, the function $U(q,p)$ does not change its functional dependence upon position and momentum, rather it changes its values as a result of replacing the values (q_A, p_A) to position (q_B, p_B) in the function $U(q,p)$.

Or, in short, $x' = x + ut$ can be understood as a Galilei transformation, or as a translation from the point (with coordinate) x to the point x'.

1.4 CONSERVED PHASE-SPACE MEASURE

Let

$$(x_1, \ldots, x_{2n}) = (p_1, \ldots, p_n, q_1, \ldots, q_n),$$
$$(X_1, \ldots, X_{2n}) = (P_1, \ldots, P_n, Q_1, \ldots, Q_n) \tag{1.37}$$

with canonical variables p_i, q_i and P_i, Q_i and $H(x) = K(X)$. Let

$$J = \begin{pmatrix} \mathbb{O} & -\mathbb{I} \\ \mathbb{I} & \mathbb{O} \end{pmatrix}, \tag{1.38}$$

---

[2] 'The traditional notation - in which one writes functions as functions of the coordinates, e.g., $H(p,q)$ - is perfectly adequate when the coordinates are fixed. On the other hand, when one changes coordinates, one has to decide whether $H(p',q')$ denotes the same function of new arguments or whether $H(p',q')$ is a different function of p' and q' which produces the same numerical value as the old function H produced with the old variables p and q. The ambiguity increases enormously when one needs to compute partial derivatives - a great deal of the complications in traditional books and papers on mechanics and thermodynamics arises from this.' (de la Llave 2001, p. 39)

with $n \times n$ block matrices $\mathbb{O}$ and $\mathbb{I}$, where $\mathbb{O}_{ij} = 0$ and $\mathbb{I}_{ij} = \delta_{ij}$. Then the Hamilton equations in the old and new variables can be written, for $\mu, \nu = 1, \ldots, 2n$,

$$\dot{x}_\mu = \sum_{\nu=1}^{2n} J_{\mu\nu} \frac{\partial H}{\partial x_\nu}, \qquad \dot{X}_\mu = \sum_{\nu=1}^{2n} J_{\mu\nu} \frac{\partial K}{\partial X_\nu}. \tag{1.39}$$

Let

$$T_{\mu\nu} = \frac{\partial X_\mu}{\partial x_\nu}. \tag{1.40}$$

Differentiating $H(x) = K(X)$ with respect to x gives, with the understanding that $X = X(x)$,

$$\frac{\partial H}{\partial x_\mu} = \sum_{\nu=1}^{2n} \frac{\partial K}{\partial X_\nu} \frac{\partial X_\nu}{\partial x_\mu}. \tag{1.41}$$

Then, for $\mu = 1, \ldots, 2n$,

$$\sum_{\tau=1}^{2n} J_{\mu\tau} \frac{\partial K}{\partial X_\tau} \overset{(1.39)}{=} \dot{X}_\mu = \sum_{\nu=1}^{2n} \frac{\partial X_\mu}{\partial x_\nu} \dot{x}_\nu = \sum_{\nu,\sigma=1}^{2n} \frac{\partial X_\mu}{\partial x_\nu} J_{\nu\sigma} \frac{\partial H}{\partial x_\sigma}$$

$$= \sum_{\nu,\sigma,\tau=1}^{2n} \frac{\partial X_\mu}{\partial x_\nu} J_{\nu\sigma} \frac{\partial X_\tau}{\partial x_\sigma} \frac{\partial K}{\partial X_\tau}$$

$$\tag{1.42}$$

Therefore,

$$\sum_{\nu,\sigma=1}^{2n} \frac{\partial X_\mu}{\partial x_\nu} J_{\nu\sigma} \frac{\partial X_\tau}{\partial x_\sigma} = J_{\mu\tau}, \tag{1.43}$$

or

$$\sum_{\nu,\sigma=1}^{2n} T_{\mu\nu} J_{\nu\sigma} T_{\tau\sigma} = J_{\mu\tau}, \tag{1.44}$$

or, with $T^\tau_{\sigma\tau} \equiv (T^\tau)_{\sigma\tau}$,

$$\sum_{\nu,\sigma=1}^{2n} T_{\mu\nu} J_{\nu\sigma} T^\tau_{\sigma\tau} = J_{\mu\tau}, \tag{1.45}$$

or, written with matrix multiplication,

$$TJT^\tau = J. \tag{1.46}$$

Taking the determinant on both sides gives

$$\det(T) \det(J) \det(T^\tau) = \det(J), \tag{1.47}$$

or, since $\det(T) = \det(T^\tau)$,

$$[\det(T)]^2 = 1, \tag{1.48}$$

or

$$\det(T) = \pm 1. \tag{1.49}$$

We consider only canonical transformations with $\det(T) = 1$. The volume/measure of a set in phase space is given by

$$\int dX_1 \ldots dX_{2n} = \int dx_1 \ldots dx_{2n} \, \det\left(\frac{\partial X_\mu}{\partial x_\nu}\right)$$

$$= \int dx_1 \ldots dx_{2n} \, \det(T) = \int dx_1 \ldots dx_{2n}, \qquad (1.50)$$

and is thus conserved by canonical transformations.

An alternative proof of $\det(\partial X_i/\partial x_j) = 1$ is given in Landau & Lifshitz (1976), par. 46. Suppressing all indices and writing $\det(A) = |A|$, one has

$$\left|\frac{\partial X}{\partial x}\right| = \left|\frac{\partial(P,Q)}{\partial(p,q)}\right| \stackrel{\text{(a)}}{=} \frac{\left|\dfrac{\partial(P,Q)}{\partial(P,q)}\right|}{\left|\dfrac{\partial(p,q)}{\partial(P,q)}\right|} \stackrel{\text{(b)}}{=} \frac{\left|\dfrac{\partial Q}{\partial q}\right|_P}{\left|\dfrac{\partial p}{\partial P}\right|_q} \stackrel{\text{(c)}}{=} 1. \qquad (1.51)$$

In (a) it is used that one can treat functional determinants like fractions, and divide (here differentiate) their nominator and denominator by the same term. If identical terms appear in the nominator and denominator of a functional determinant, they can be assumed to be constant (indicated by subscripts 'P' and 'q') and crossed out, as is done in (b). In (c), the last equation in (1.55) below is used.

1.5 COORDINATE RELATIONS

Writing $TJT^\tau = J$ as $JT^\tau = T^{-1}J$ (coordinate transformations are invertible) leads to another characterization of canonical transformations. Instead of using J and T, we use again p_i, q_i and P_i, Q_i. Then, for $i, j = 1, \ldots, n$,

$$\dot{P}_i = \sum_j \frac{\partial P_i}{\partial p_j} \dot{p}_j + \sum_j \frac{\partial P_i}{\partial q_j} \dot{q}_j = -\sum_j \frac{\partial P_i}{\partial p_j} \frac{\partial H}{\partial q_j} + \sum_j \frac{\partial P_i}{\partial q_j} \frac{\partial H}{\partial p_j}. \qquad (1.52)$$

Alternatively, using $p_i = p_i(P,Q)$ and $q_i = q_i(P,Q)$ and assuming $H(p,q) = K(P,Q)$,

$$\dot{P}_i = -\frac{\partial K}{\partial Q_i} = -\sum_j \frac{\partial H}{\partial p_j} \frac{\partial p_j}{\partial Q_i} - \sum_j \frac{\partial H}{\partial q_j} \frac{\partial q_j}{\partial Q_i}. \qquad (1.53)$$

Comparing Eqs. (1.52) and (1.53) gives, for $i, j = 1, \ldots, n$,

$$\frac{\partial P_i}{\partial p_j} = \frac{\partial q_j}{\partial Q_i} \qquad \text{and} \qquad \frac{\partial P_i}{\partial q_j} = -\frac{\partial p_j}{\partial Q_i}. \qquad (1.54)$$

Similarly, from the $\dot{Q}_i$ one obtains

$$\frac{\partial Q_i}{\partial p_j} = -\frac{\partial q_j}{\partial P_i} \qquad \text{and} \qquad \frac{\partial Q_i}{\partial q_j} = \frac{\partial p_j}{\partial P_i}. \qquad (1.55)$$

Equations (1.54) and (1.55) occur in Jacobi (1884), lecture 31, and must be obeyed in a canonical transformation $p_i, q_i \to P_i, Q_i$. We check that this is the case for the transformation in Eq. (1.23). From $p_j = S_{q_j}$ with $S = S(P, q)$ it follows that

$$\frac{\partial p_j}{\partial P_i} = S_{q_j P_i}, \tag{1.56}$$

and from $Q_i = S_{P_i}$,

$$\frac{\partial Q_i}{\partial q_j} = S_{P_i q_j}. \tag{1.57}$$

Equations (1.56) and (1.57) give the second equation in (1.55); their inverses give the first equation in (1.54). Furthermore,

$$\frac{\partial p_j}{\partial Q_i} = \frac{\partial}{\partial Q_i} \frac{\partial S(P, q)}{\partial q_j} = 0, \tag{1.58}$$

and

$$\frac{\partial Q_i}{\partial p_j} = \frac{\partial}{\partial p_j} \frac{\partial S(P, q)}{\partial P_i} = 0. \tag{1.59}$$

Finally, since $S = S(P, q)$ with independent variables P_i, q_j,

$$\frac{\partial P_i}{\partial q_j} = \frac{\partial q_j}{\partial P_i} = 0. \tag{1.60}$$

Equations (1.58) to (1.60) establish the second equation in (1.54) and the first in (1.55), thus Eq. (1.23) is a canonical transformation.

1.6 ACTION-ANGLE VARIABLES

Action-angle variables[3] J_i, φ_i are introduced by demanding that the Hamiltonian K after a canonical transformation depends on the action variables $J_1, \ldots, J_n$ alone,

$$K = K(J). \tag{1.61}$$

The Hamilton equations become then

$$\frac{dJ_i}{dt} = \frac{\partial K}{\partial \varphi_i} = 0, \qquad \frac{d\varphi_i}{dt} = \frac{\partial K}{\partial J_i}. \tag{1.62}$$

Thus J_i is constant, and so is $\dot{\varphi}_i$ as derivative of K at fixed J_i,

$$\dot{\varphi}_i = \omega_i = \text{const.} \tag{1.63}$$

---

[3]The action variable J_i is not to be confused with the matrix J from Eq. (1.38).

The Hamilton equations are therefore solved as

$$J_i = a_i, \qquad \varphi_i = \omega_i t + \varphi_0, \qquad (1.64)$$

where a_i, ω_i, φ_0 are $3n$ constants. Thus, if n constant momenta J_i of a system are known, the Hamilton equations are completely solved! The Liouville theorem of the next section generalizes this statement.

Systems with n degrees of freedom for which a canonical transformation to n pairs of action-angle variables is possible are called (complete) *integrable*. Many or all mechanical systems with a known analytic solution are integrable. The standard example is the *harmonic oscillator*, with Hamilton function

$$H = \tfrac{1}{2}\left(p^2 + q^2\right), \qquad (1.65)$$

where the frequency is normalized to unity. The Hamilton equations are

$$\dot{p} = -q, \qquad \dot{q} = p, \qquad (1.66)$$

or

$$\frac{\dot{p}}{\dot{q}} = \frac{dp}{dq} = -\frac{q}{p} \qquad (1.67)$$

or

$$p\,dp + q\,dq = 0, \qquad (1.68)$$

with solution

$$p^2 + q^2 = 2e = \text{const}, \qquad (1.69)$$

where e is the kinetic plus potential energy (per mass) of the oscillator. The trajectories in phase space are therefore circles. Let $J = e$, then $K = J$ and $\dot{\varphi} = 1$ or $\varphi = t$. Then J, φ are action-angle variables, where J parameterizes the circles in phase space, and φ is the polar angle on the circle(s).

These results carry over to n independent harmonic oscillators. Each of them carries its own pair J_i, φ_i of action-angle variables, and the trajectory lies on an n-dimensional torus in $2n$-dimensional phase space, the tori are nested inside each other.

If all $n(n-1)/2$ frequency ratios ω_i/ω_j are rational, the trajectory is closed; if some frequency ratios are irrational ('incommensurable'), which means that 'the ω_i are linearly independent over $\mathbb{Z}$ (i.e., no non-trivial linear combination $\sum k_i \omega_i$ with integers k_i is equal to zero,' Anosov 1995, p. 218), the trajectory fills the torus densely, which is called a quasi-periodic motion:

Definition (e.g., Rüssmann 1979, p. 203)   A function (especially, a trajectory) $f(t) : \mathbb{R} \to \mathbb{R}^m$ with $m \geq 1$ is called *quasi-periodic* with n frequencies $\omega_1, \ldots, \omega_n$ if (a) there is a continuous function $F : \mathbb{R}^n \to \mathbb{R}^m$ such that

$$f(t) = F(\omega_1 t, \ldots, \omega_n t), \qquad (1.70)$$

(b) F is 2π-periodic in each of its n arguments and (c) the ω_i are linearly independent over $\mathbb{Z}$,

$$k_1\omega_1 + \cdots + k_n\omega_n = 0 \quad \rightarrow \quad k_1 = \cdots = k_n = 0 \qquad (1.71)$$

with $k_i \in \mathbb{Z}$ for $i = 1,\ldots,n$.

Thus phase-space tori are central to integrable systems (that are bounded in space), and the KAM theorem is about their stability under small perturbations. Somewhat strangely, most classical texts on Hamilton mechanics do not mention phase-space tori.

1.7   INTEGRABLE SYSTEMS AND PHASE-SPACE TORI

We start with some quotations that capture the essence of this section.

> A Hamiltonian system with $2n$-dimensional phase space is *integrable* if there exist n independent, mutually commuting smooth functions $F_1,\ldots,F_n$ [...] [T]he F_k are *integrals*, i.e. constants of the motion along phase-space orbits. (Lowenstein 2012, p. 56)

> In the nineteenth century, the general and very useful notion of *Liouville integrability* for Hamiltonian systems, was introduced: If a Hamiltonian system with Hamiltonian H and n degrees of freedom has n independent, Poisson-commuting integrals $I_1,\ldots,I_n$, then the flow $t \mapsto z(t)$ generated by H can be integrated explicitly by quadrature. (Deift 2019, p. 3)

> There is a lot of controversy over which of the known integrability mechanisms is most fundamental, but there is a consensus that integrability means a complete set of Poisson-commuting first integrals. This definition and 'Liouville's Theorem' on geometric consequences of the integrability property (namely, foliation of the phase space by Lagrangian tori) are in fact Arnold's original inventions. (Khesin & Tabachnikov 2012, p. 386)

In this section, Liouville's theorem on integrable systems is proved. This theorem was substantially extended by Arnold (1963c), who showed that trajectories of integrable systems lie on surfaces of n-dimensional tori $T = S \times \cdots \times S$ in $2n$-dimensional phase space (S is the 1-circle and $\times$ the topological product, which is the Cartesian product of sets equipped with the product topology). The following proof of the classical part of the Liouville theorem is taken from Whittaker (1917). Some mathematical background on Liouville's theorem using the *reduction procedure* of E. Cartan on symplectic manifolds can be found in Abraham & Marsden (1978), pp. 298-301. At the end of this section, Arnold's extension of Liouville's theorem is discussed.

Let $y_1,\ldots,y_n$ be independent variables in $\mathbb{R}^n$ (in the following, the canonical variables p_i, q_i). A *differential one-form* or *Pfaff form* is any expression

$$Y_1(y_1,\ldots,y_n)\,\mathrm{d}y_1 + \cdots + Y_n(y_1,\ldots,y_n)\,\mathrm{d}y_n, \qquad (1.72)$$

where the Y_i are twice-continuously differentiable functions. One considers these forms as elements of a *cotangent* vector space $T^* \mathbb{R}^n$ with basis vectors[4] $\mathrm{d}y_j$. Let $f = (f_1, \ldots, f_n) : \mathbb{R}^m \to \mathbb{R}^n$ be a continuously differentiable function,

$$y_j = f_j(x_1, \ldots, x_m) \tag{1.73}$$

with $j = 1, \ldots, n$. Then

$$\mathrm{d}y_j = \sum_{i=1}^{n} \frac{\partial y_j}{\partial x_i} \, \mathrm{d}x_i, \tag{1.74}$$

and the form in Eq. (1.72) is written in the new variables x_i as

$$\sum_{i=1}^{n} X_i(x) \, \mathrm{d}x_i = \sum_{j=1}^{n} Y_j(y) \, \mathrm{d}y_j \tag{1.75}$$

with

$$X_i(x) = \sum_{j=1}^{n} \frac{\partial y_j}{\partial x_i} Y_j(y(x)). \tag{1.76}$$

The chain rule gives for the gradients,

$$\frac{\partial}{\partial x_i} = \sum_{j=1}^{n} \frac{\partial y_j}{\partial x_i} \frac{\partial}{\partial y_j}. \tag{1.77}$$

The $\partial/\partial x_i$ and $\partial/\partial y_j$ can be used as basis vectors of the vector spaces $\mathbb{R}^m$ and $\mathbb{R}^n$ equipped with coordinates x_i and y_j, respectively. Equation (1.74) is the transformation law of ('covariant') forms $\mathrm{d}x_i$, and Eq. (1.76), according to Eq. (1.77), that of ('contravariant') vectors. Equation (1.75) can be read as (scalar) invariance of 1-forms under a change of variables,

$$\sum_i Y_i \, \mathrm{d}y_i = \sum_{i,j,k} \frac{\partial x_j}{\partial y_i} \frac{\partial y_i}{\partial x_k} X_j \, \mathrm{d}x_k = \sum_{j,k} \frac{\partial x_j}{\partial x_k} X_j \, \mathrm{d}x_k = \sum_{j,k} \delta_{jk} X_j \, \mathrm{d}x_k = \sum_j X_j \, \mathrm{d}x_j. \tag{1.78}$$

Traditional Ricci calculus works with the transformation matrix $(\partial y_j / \partial x_i)$ and its inverse. The modern approach is to use instead Eq. (1.75) as definition of the *pullback* of a form, $f^* : T^* \mathbb{R}^n \to T^* \mathbb{R}^m$,

$$\boxed{f^* \left(\sum_j Y_j \, \mathrm{d}y_j \right) = \sum_j Y_j \circ f \, \mathrm{d}f_j} \tag{1.79}$$

where, as in Eq. (1.74),

$$\mathrm{d}f_j = \sum_i \frac{\partial f_j}{\partial x_i} \, \mathrm{d}x_i. \tag{1.80}$$

---

[4]The $\mathrm{d}y_j$ are here *not* infinitesimal quantities, but have finite magnitude: they are the differentials ('linearization') of the coordinate functions $y_j : (y_1, \ldots, y_j, \ldots, y_n) \mapsto y_j$. Some books therefore use $\mathrm{D}y_j$ instead of $\mathrm{d}y_j$.

For Eq. (1.79) and its obvious generalization to k-forms, see Bott & Tu (1982), p. 19, or Forster (2017), p. 256, or almost any book on differential geometry. In matrix notation, Eq. (1.76) can be written as (see Forster 2017, p. 257)

$$X = Y \circ f \ Df, \qquad (1.81)$$

where $Df = (\partial f_i/\partial x_j)$ is the functional matrix or derivative of f.

Since the X_i and $\partial/\partial x_i$ transform identically, but the $\mathrm{d}x_i$ with the inverse matrix, one obtains

$$\sum_{i,j} \left(\frac{\partial X_i}{\partial x_j} - \frac{\partial X_j}{\partial x_i} \right) \mathrm{d}x_i \, \delta x_j = \sum_{k,l} \left(\frac{\partial Y_k}{\partial y_l} - \frac{\partial Y_l}{\partial y_k} \right) \mathrm{d}y_k \, \delta y_l. \qquad (1.82)$$

Here $\mathrm{d}x_i$ and δx_i are two independent differentials of the x_i (correspondingly for $\mathrm{d}y_i$ and δy_i). The reason for introducing a new δx_i is that $a_{ij} \, \mathrm{d}x_i \, \mathrm{d}x_j$ with $a_{ij} = -a_{ji}$ as in Eq. (1.82) would vanish identically (antisymmetric times symmetric matrix), whereas

$$\sum_{i,j} \left(\frac{\partial X_i}{\partial x_j} - \frac{\partial X_j}{\partial x_i} \right) \mathrm{d}x_i \, \delta x_j = \frac{1}{2} \sum_{i,j} \left(\frac{\partial X_i}{\partial x_j} - \frac{\partial X_j}{\partial x_i} \right) (\mathrm{d}x_i \, \delta x_j - \mathrm{d}x_j \, \delta x_i) \neq 0. \quad (1.83)$$

According to Giaquinta & Hildebrandt (2004), p. 438, the differential of a Pfaff form using two differentials $\mathrm{d}x_i$ and δx_i as in Eq. (1.82) occured for the first time in Schering (1873). The modern approach is to use the Cartan alternating product[5] $\mathrm{d}x_i \wedge \mathrm{d}x_j$.

Poincaré invariant   The most important differential 1- and 2-forms of classical mechanics occur in the Poincaré invariant J_1,

$$J_1 = \oint_{\partial A} \sum_i p_i \, \mathrm{d}q_i = \iint_A \sum_i \mathrm{d}p_i \, \mathrm{d}q_i, \qquad (1.84)$$

where A is an arbitrary two-dimensional surface in phase space with boundary ∂A. In the second equation, Stokes' theorem is used. J_1 is invariant under canonical transformations. For a proof with line integral $\oint_{\partial A}$, see Lanczos (1970), p. 181; for a proof with surface integral $\iint_A$, see Nordheim & Fues (1927), pp. 102-103, and Lanczos (1970), pp. 210-212.

Choosing as surface A in Eq. (1.84) an infinitesimal parallelogram with one pair of sides given by the vector[6] $(\mathrm{d}p_1, \ldots, \mathrm{d}p_n, \mathrm{d}q_1, \ldots, \mathrm{d}q_n)$ and the other pair of sides by $(\delta p_1, \ldots, \delta p_n, \delta q_1, \ldots, \delta q_n)$ (arbitrary tilt in phase space), the sum in Eq. (1.84) is over the areas of the projections of the parallelogram into the n planes (p_i, q_i),

---

[5]The $\mathrm{d}x_i$ and δx_i are linear functions acting on vectors. In the product $\mathrm{d}x_i \, \delta x_j$, the order of the factors is relevant, as in $\mathrm{d}x_i \wedge \mathrm{d}x_j$. In physics one writes instead $\mathrm{d}x_i \, \mathrm{d}x_j'$ for the product of two independent infinitesimal increments.

[6]We use here a mixture of the mathematical and physical meaning of differentials.

giving again parallelograms. The i-th parallelogram area is given by the cross product $(\mathrm{d}p_i, \mathrm{d}q_i) \times (\delta p_i, \delta q_i) = \mathrm{d}p_i\, \delta q_i - \delta p_i\, \mathrm{d}q_i$, thus the quantity

$$\omega = \sum_i (\mathrm{d}p_i\, \delta q_i - \delta p_i\, \mathrm{d}q_i) \tag{1.85}$$

is an invariant under canonical transformations, termed the *bilinear invariant* (see Nordheim & Fues 1927, p. 104) or *symplectic two-form*. With the alternating product, this becomes $\omega = \sum_i \mathrm{d}p_i \wedge \mathrm{d}q_i$. Note that ω depends only on the coordinate functions p_i and q_i. This skew-symmetric 2-form ω is as important for symplectic manifolds as is the symmetric rank-2 metric tensor g for Riemann manifolds: it establishes an isomorphism between 1-forms and vectors.

Definition An *exact* 1-form is the total differential of a function,

$$\sum_i X_i(x)\, \mathrm{d}x_i = \mathrm{d}F(x) = \sum_i \frac{\partial F}{\partial x_i}\, \mathrm{d}x_i. \tag{1.86}$$

For an exact differential 1-form, the 2-form in Eq. (1.82) vanishes, since

$$\sum_{i,j} \left(\frac{\partial X_i}{\partial x_j} - \frac{\partial X_j}{\partial x_i} \right) = \sum_{i,j} \left(\frac{\partial^2 F}{\partial x_j\, \partial x_i} - \frac{\partial^2 F}{\partial x_i\, \partial x_j} \right) = 0. \tag{1.87}$$

Under certain restrictions, the opposite statement is also true: if Eq. (1.87) holds for the X_i, then the form $X_i\, \mathrm{d}x_i$ is exact. The most general form of this statement, that any *closed* differential form on a contractible domain is exact, is termed Poincaré's theorem (or lemma). The proof is somewhat technical and can be found in Flanders (1989), pp. 27-29, and Bott & Tu (1982), pp. 33-35. We consider here only Pfaff 1-forms on *star domains*.[7]

Definition A domain U in $\mathbb{R}^n$ is a *star domain* if there is a point $x_0 \in U$ so that for all $y \in U$ the straight segment from x to y is in U. Every convex set is a star domain, but the opposite is not true; convexity is thus the stronger condition. Every star domain is contractible, thus contractibility is a weaker condition. Furthermore, any star domain is a simply connected[8] set. Star domains play a central role in complex function theory, in proving (a simple version of) the Cauchy theorem.

The following lemma and proof are taken from Fritzsche (2011), see also Forster (2017), pp. 233-234.

Lemma 2 *Let U be a star domain in $\mathbb{R}^n$. If a Pfaffian $\sum_i X_i\, \mathrm{d}x_i$ on U obeys*

$$\frac{\partial X_i}{\partial x_j} = \frac{\partial X_j}{\partial x_i} \tag{1.88}$$

---

[7]Whittaker (1917) applies Poincaré's lemma without proving it.

[8]In two dimensions, this is a domain without holes; in $n > 2$ dimensions, it is a domain without 'tunnels' from boundary to boundary, as in a torus = ball minus central cylinder.

for all i, j, then $\sum_i X_i \, dx_i$ is exact, $\sum_i X_i \, dx_i = dF$.

Proof Let $x_0 = 0$ and

$$F(x) = \sum_i x_i \int_0^1 dt \, X_i(tx), \qquad (1.89)$$

where the sum over all i is understood. The integral on the right side of Eq. (1.89) is well-defined, i.e., can be evaluated for all $x \in U$ since U is a star domain. Then, using an auxiliary variable $y_i = tx_i$ with

$$\frac{\partial}{\partial x_i} = t \frac{\partial}{\partial y_i}, \qquad (1.90)$$

one obtains

$$\begin{aligned}
\frac{\partial F}{\partial x_j} &\overset{(1.89)}{=} \sum_i \delta_{ij} \int_0^1 dt \, X_i(tx) + \sum_i x_i \frac{\partial}{\partial x_j} \int_0^1 dt \, X_i(tx) \\
&\overset{(1.88)}{=} \int_0^1 dt \left[X_j(tx) + \sum_i \frac{\partial X_j}{\partial x_i}(tx) \frac{d(tx_i)}{dt} \right] \\
&\overset{(1.90)}{=} \int_0^1 dt \left[X_j(tx) + t \sum_i \frac{\partial X_j}{\partial y_i}(y) \frac{dy_i}{dt} \right] \\
&= \int_0^1 dt \left[X_j(tx) + t \frac{dX_j}{dt}(tx) \right] \\
&= \int_0^1 dt \, \frac{d}{dt}(tX_j(tx)) \\
&= tX_j(tx)\big|_0^1 \\
&= X_j(x). \qquad (1.91)
\end{aligned}$$

Therefore, $\sum_j X_j \, dx_j = \sum_j \frac{\partial F}{\partial x_j} dx_j = dF$. $\qquad \square$

The following simple observation is important.

Lemma 3 *Let X_i and Y_j be related as in Eq. (1.76) and let $\partial/\partial x_i$ and $\partial/\partial y_j$ be as in Eq. (1.77). Let*

$$a_{ij} = \left(\frac{\partial X_i}{\partial x_j} - \frac{\partial X_j}{\partial x_i} \right) \qquad and \qquad b_{ij} = \left(\frac{\partial Y_i}{\partial y_j} - \frac{\partial Y_j}{\partial y_i} \right). \qquad (1.92)$$

Then $\sum_j a_{ij} \, dx_j = 0$ implies $\sum_j b_{ij} \, dy_j = 0$.

Proof

$$\sum_j b_{ij}\, \mathrm{d}y_j = \sum_j \left(\frac{\partial Y_i}{\partial y_j} - \frac{\partial Y_j}{\partial y_i} \right) \mathrm{d}y_j$$

$$= \sum_{j,k,l,m} \left(\frac{\partial x_l}{\partial y_j} \frac{\partial x_k}{\partial y_i} \frac{\partial X_k}{\partial x_l} - \frac{\partial x_l}{\partial y_i} \frac{\partial x_k}{\partial y_j} \frac{\partial X_k}{\partial x_l} \right) \frac{\partial y_j}{\partial x_m}\, \mathrm{d}x_m$$

$$= \sum_{k,l,m} \left(\delta_{lm} \frac{\partial x_k}{\partial y_i} \frac{\partial X_k}{\partial x_l} - \delta_{km} \frac{\partial x_l}{\partial y_i} \frac{\partial X_k}{\partial x_l} \right) \mathrm{d}x_m$$

$$= \sum_{k,l} \frac{\partial x_k}{\partial y_i} \frac{\partial X_k}{\partial x_l}\, \mathrm{d}x_l - \sum_{k,l} \frac{\partial x_l}{\partial y_i} \frac{\partial X_k}{\partial x_l}\, \mathrm{d}x_k$$

$$= \sum_{k,l} \frac{\partial x_k}{\partial y_i} \left(\frac{\partial X_k}{\partial x_l} - \frac{\partial X_l}{\partial x_k} \right) \mathrm{d}x_l$$

$$= \sum_{k,l} \frac{\partial x_k}{\partial y_i} a_{kl}\, \mathrm{d}x_l$$

$$= 0. \tag{1.93}$$

Definition For n independent variables x_i, the n equations

$$\sum_{j=1}^{n} \left(\frac{\partial X_j}{\partial x_i} - \frac{\partial X_i}{\partial x_j} \right) \mathrm{d}x_j = 0 \tag{1.94}$$

for $i = 1,\ldots,n$ are called the *first Pfaff system* of the 1-form $\sum_i X_i\, \mathrm{d}x_i$.

Let p_i, q_i be canonical variables for a system with Hamiltonian $H(p,q,t)$.

Lemma 4 *The first Pfaff system of the 1-form*

$$\sum_{i=1}^{n} p_i\, \mathrm{d}q_i - H\, \mathrm{d}t \tag{1.95}$$

are the Hamilton equations.

Proof Introduce two vectors with $2n+1$ components each,

$$x = (p_1,\ldots,p_n,q_1,\ldots,q_n,t),$$
$$X = (0,\ldots,0,p_1,\ldots,p_n,-H). \tag{1.96}$$

Then

$$\sum_{i=1}^{n} p_i\, \mathrm{d}q_i - H\, \mathrm{d}t = \sum_{\mu=1}^{2n+1} X_\mu\, \mathrm{d}x_\mu. \tag{1.97}$$

Direct calculation of the first Pfaff system of this 1-form gives, with $i, j = 1, \ldots, n$,

$$a_{ij} = 0 - 0 = 0,$$

$$a_{n+i,j} = \frac{\partial p_i}{\partial p_j} - 0 = \delta_{ij},$$

$$a_{i,n+j} = 0 - \frac{\partial p_j}{\partial p_i} = -\delta_{ij},$$

$$a_{n+i,n+j} \overset{(\P)}{=} \frac{\partial p_i}{\partial q_j} - \frac{\partial p_j}{\partial q_i} = 0,$$

$$a_{i,2n+1} = 0 + \frac{\partial H}{\partial p_i} = \frac{\partial H}{\partial p_i},$$

$$a_{2n+1,j} = -\frac{\partial H}{\partial p_j} - 0 = -\frac{\partial H}{\partial p_j},$$

$$a_{n+i,2n+1} \overset{(\P)}{=} \frac{\partial p_i}{\partial t} + \frac{\partial H}{\partial q_i} = \frac{\partial H}{\partial q_i},$$

$$a_{2n+1,n+j} \overset{(\P)}{=} -\frac{\partial H}{\partial q_j} - \frac{\partial p_j}{\partial t} = -\frac{\partial H}{\partial q_j},$$

$$a_{2n+1,2n+1} = -\frac{\partial H}{\partial t} + \frac{\partial H}{\partial t} = 0. \tag{1.98}$$

In $(\P)$ we used that p_i, q_i, t are independent variables. In matrix notation, the first Pfaff system becomes (with n-vector and $(n \times n)$-matrix)

$$
\begin{pmatrix}
0 & -\delta_{ij} & \dfrac{\partial H}{\partial p_i} \\[2ex]
\delta_{ij} & 0 & \dfrac{\partial H}{\partial q_i} \\[2ex]
-\dfrac{\partial H}{\partial p_j} & -\dfrac{\partial H}{\partial q_j} & 0
\end{pmatrix}
\begin{pmatrix}
dp_j \\[2ex]
dq_j \\[2ex]
dt
\end{pmatrix}
=
\begin{pmatrix}
0 \\[2ex]
0 \\[2ex]
0
\end{pmatrix}. \tag{1.99}
$$

Written component-wise $(i = 1, \ldots, n)$,

$$-dq_i + \frac{\partial H}{\partial p_i} \, dt = 0, \tag{1.100}$$

$$dp_i + \frac{\partial H}{\partial q_i} \, dt = 0, \tag{1.101}$$

$$\sum_{j=1}^{n} \left(\frac{\partial H}{\partial p_j} \, dp_j + \frac{\partial H}{\partial q_j} \, dq_j \right) = 0, \tag{1.102}$$

which are the Hamilton equations,

$$\frac{\mathrm{d}q_i}{\mathrm{d}t} = \frac{\partial H}{\partial p_i}, \tag{1.103}$$

$$\frac{\mathrm{d}p_i}{\mathrm{d}t} = -\frac{\partial H}{\partial q_i}, \tag{1.104}$$

$$\frac{\mathrm{d}H}{\mathrm{d}t} = \frac{\partial H}{\partial t}, \tag{1.105}$$

where we added $-(\partial H/\partial t)\,\mathrm{d}t$ to the left and right side of Eq. (1.102) to obtain Eq. (1.105). The latter actually follows from Eqs. (1.103) and (1.104),

$$\frac{\mathrm{d}H}{\mathrm{d}t} = \frac{\partial H}{\partial t} + \frac{\partial H}{\partial p_i}\frac{\mathrm{d}p_i}{\mathrm{d}t} + \frac{\partial H}{\partial q_i}\frac{\mathrm{d}q_i}{\mathrm{d}t} = \frac{\partial H}{\partial t} - \frac{\partial H}{\partial p_i}\frac{\partial H}{\partial q_i} + \frac{\partial H}{\partial q_i}\frac{\partial H}{\partial p_i} = \frac{\partial H}{\partial t}, \tag{1.106}$$

and thus contains no extra information. □

Definition Let $u_1(p,q),\ldots,u_r(p,q)$ be $r \leq n$ functions of $2n$ canonical variables p_i, q_i. The u_i are said to be *in involution* if

$$\{u_i, u_j\} = 0 \tag{1.107}$$

for all i, j from 1 to r, where $\{.,.\}$ is the Poisson bracket introduced in Eq. (1.34).

Lemma 5 *Let $u_1(p,q),\ldots,u_r(p,q)$ be in involution. If*

$$u_1(p,q) = \cdots = u_r(p,q) = 0 \tag{1.108}$$

and if

$$u_1(p,q) = \cdots = u_r(p,q) = 0 \;\; \rightarrow \;\; v(p,q) = 0, \tag{1.109}$$

then $\{u_1, v\} = \cdots = \{u_r, v\} = 0$.

Proof Let m with $1 \leq m \leq r$ be arbitrary but fixed. For $i = 1,\ldots,n$, let

$$P_i = p_i - \varepsilon\,\frac{\partial u_m}{\partial q_i},$$

$$Q_i = q_i + \varepsilon\,\frac{\partial u_m}{\partial p_i} \tag{1.110}$$

be infinitesimal changes in p_i and q_i as in Eq. (1.33), with K replaced by u_m. Then from Eq. (1.35)

$$\mathrm{d}u_i = u_i(P,Q) - u_i(p,q) = \varepsilon\,\{u_m, u_i\} = 0, \tag{1.111}$$

according to the assumption of the lemma. Note that $u_i(P,Q)$ appears here, but no new function $U_i(P,Q)$. Thus, if $u_i(p,q) = 0$ then $u_i(P,Q) = 0$ for $i = 1,\ldots,r$. If $u_1(p,q) = \cdots = u_r(p,q) = 0$ implies $v(p,q) = 0$, then $u_1(P,Q) = \cdots = u_r(P,Q) = 0$ implies $v(P,Q) = 0$ by renaming the variables. Thus

$$\mathrm{d}v = v(P,Q) - v(p,q) = 0 - 0 = 0. \tag{1.112}$$

According to Eq. (1.35),

$$dv = \varepsilon \{u_m, v\},$$ (1.113)

and thus

$$\{u_m, v\} = 0.$$ (1.114)

Since m was arbitrary, it follows that $\{u_1, v\} = \cdots = \{u_r, v\} = 0$. □

Lemma 6 *Let $u_1(p,q), \ldots, u_r(p,q)$ be in involution and $u_1(p,q) = \cdots = u_r(p,q) = 0$ imply $v(p,q) = w(p,q) = 0$. Then $\{v, w\} = 0$.*

Proof From Lemma 5, $\{u_1, v\} = \cdots = \{u_r, v\} = 0$, thus $u_1, \ldots, u_r, v$ are in involution. If $u_1 = \cdots = u_r = 0$ implies $w = 0$, so does $u_1 = \cdots = u_r = v = 0$. Thus the conditions of Lemma 5 are (again) met, and the lemma gives $\{u_1, w\} = \cdots = \{u_r, w\} = \{v, w\} = 0$. The last equation was to be shown. □

The next theorem shows that if n constants of motion of a Hamiltonian system with n degrees of freedom are known, the problem of solving the $2n$ differential equations of first order is essentially already solved: it remains only to perform n integrations, traditionally called quadratures. In the following equations, subscripts i and j are assumed to run from 1 to n.

Theorem (Liouville) *Let*

$$\frac{dq_i}{dt} = \frac{\partial H}{\partial p_i}, \qquad \frac{dp_i}{dt} = -\frac{\partial H}{\partial q_i}$$ (1.115)

be a Hamiltonian system with $H = H(p_1, \ldots, p_n, q_1, \ldots, q_n, t)$ and let

$$I_i(p_1, \ldots, p_n, q_1, \ldots, q_n, t) = a_i$$ (1.116)

with $i = 1, \ldots, n$ be n mutually independent constants of motion,[9] with arbitrary real constants $a_1, \ldots, a_n$, so that one can obtain the momenta p_i as functions [10]

$$p_i = P_i(a_1, \ldots, a_n, q_1, \ldots, q_n, t).$$ (1.117)

The functions I_1 to I_n shall be in involution,

$$\{I_i, I_j\} = 0.$$ (1.118)

(I) *Then the Pfaffian 1-form $p_1 \, dq_1 + \cdots + p_n \, dq_n - H \, dt$ that characterizes the Hamiltonian system can be written as an exact differential $dV(a_1, \ldots, a_n, q_1, \ldots, q_n, t)$.*

---

[9] I_i is a constant of motion if $dI_i/dt = 0$. Thus, the function I_i may actually depend on time t as independent variable, $\partial I_i/\partial t \neq 0$. If instead $\partial I_i/\partial t = 0$, then I_i is called *stationary*.

[10] Arnold (1989), pp. 272-273, assumes that the I_i are independent in the sense that their gradients (or the 1-forms dI_i) are *linearly* independent. The existence of functions P_i as in Eq. (1.117) follows then from the implicit function theorem.

(II) *The equations of motion become*

$$\frac{\partial V}{\partial a_i} = b_i, \tag{1.119}$$

with arbitrary real constants b_1 to b_n.

Proof (see Whittaker 1917, par. 148)

(I) Since the I_i are in involution, so are the functions $I_i - a_i$, which all vanish. Since $I_1 - a_1 = \cdots = I_n - a_n = 0$ implies $p_1 - P_1 = \cdots = p_n - P_n = 0$, one has by Lemmas 5 and 6 that the $p_i - P_i$ are in involution too,

$$\{p_i - P_i, p_j - P_j\} = 0. \tag{1.120}$$

By definition of the Poisson bracket, this gives the integrability conditions

$$\frac{\partial P_j}{\partial q_i} = \frac{\partial P_i}{\partial q_j}. \tag{1.121}$$

Next, integrability conditions containing $\partial P_i/\partial t$ are derived. One has, for constant a_1 to a_n,

$$-\frac{\partial H}{\partial q_i} = \frac{dp_i}{dt} = \frac{dP_i}{dt} = \frac{\partial P_i}{\partial t} + \sum_{j=1}^{n} \frac{\partial P_i}{\partial q_j} \frac{dq_j}{dt} =$$
$$= \frac{\partial P_i}{\partial t} + \sum_{j=1}^{n} \frac{\partial P_i}{\partial q_j} \frac{\partial H}{\partial p_j} = \frac{\partial P_i}{\partial t} + \sum_{j=1}^{n} \frac{\partial P_j}{\partial q_i} \frac{\partial H}{\partial p_j}, \tag{1.122}$$

and therefore

$$\frac{\partial P_i}{\partial t} = -\frac{\partial H}{\partial q_i} - \sum_{j=1}^{n} \frac{\partial H}{\partial p_j} \frac{\partial P_j}{\partial q_i}. \tag{1.123}$$

Let

$$K(a_1,\ldots,a_n,q_1,\ldots,q_n,t) = \tag{1.124}$$
$$H(P_1(a_1,\ldots,a_n,q_1,\ldots,q_n,t),\ldots,P_n(a_1,\ldots,a_n,q_1,\ldots,q_n,t),q_1,\ldots,q_n,t).$$

This gives, with $H = H(p_1,\ldots,p_n,q_1,\ldots,q_n,t)$,

$$\frac{\partial K}{\partial q_i} = \sum_{j=1}^{n} \frac{\partial H}{\partial p_j} \frac{\partial P_j}{\partial q_i} + \frac{\partial H}{\partial q_i}, \tag{1.125}$$

and therefore by Eq. (1.123),

$$\frac{\partial P_i}{\partial t} = -\frac{\partial K}{\partial q_i}. \tag{1.126}$$

We assume that the $(q_1,\ldots,q_n,t)$ belong to a star domain. Then the integrability conditions (1.121) and (1.126) imply[11] by Lemma 2 (see Eq. 1.88) that for constant a_i,

$$\sum_{i=1}^{n} P_i(a,q,t)\,\mathrm{d}q_i - K(a,q,t)\,\mathrm{d}t = \mathrm{d}V(a,q,t), \tag{1.127}$$

i.e., the 1-form on the left is exact.

(II) In the second part of the proof, the a_i are treated as variables. A set of constant a_i refers to a submanifold in phase space to which the motion is restricted; a change in the a_i is then a 'motion' (parameter change) from one of these submanifolds to another. From Eq. (1.127), allowing for changes in the a_i,

$$\mathrm{d}V(a,q,t) = \sum_{i=1}^{n} \frac{\partial V}{\partial a_i}\,\mathrm{d}a_i + \sum_{i=1}^{n} P_i(a,q,t)\,\mathrm{d}q_i - K(a,q,t)\,\mathrm{d}t. \tag{1.128}$$

This is rewritten as

$$\mathrm{d}V(a,q,t) - \sum_{i=1}^{n} \frac{\partial V}{\partial a_i}\,\mathrm{d}a_i = \sum_{i=1}^{n} P_i(a,q,t)\,\mathrm{d}q_i - K(a,q,t)\,\mathrm{d}t. \tag{1.129}$$

The right side is the Hamiltonian 1-form $\sum_i p_i\,\mathrm{d}q_i - H\,\mathrm{d}t$ for constant a_i. We use Lemma 4 to obtain the Hamilton equations in the variables a_i,q_i,t from the first Pfaff system of the 1-form on the left side of Eq. (1.129). The first Pfaff system of the exact differential form $\mathrm{d}V(a,q,t)$ is trivial, $0 = 0$, since second derivatives of V commute. Thus we need only consider $-\sum_i (\partial V/\partial a_i)\,\mathrm{d}a_i$. Introducing again two vectors x and X with $2n+1$ components,

$$x = (a_1,\ldots,a_n,q_1,\ldots,q_n,t), \tag{1.130}$$

$$X = \left(-\frac{\partial V}{\partial a_1},\ldots,-\frac{\partial V}{\partial a_n},0,\ldots,0,0\right), \tag{1.131}$$

one obtains for $b_{\mu\nu} = \partial X_\mu/\partial x_\nu - \partial X_\nu/\partial x_\mu$ with $\mu,\nu = 1,\ldots,2n+1$, then again for

---

[11] The integrability condition including $\partial K/\partial t$ is trivial, $\partial K/\partial t = \partial K/\partial t$, since here $i = j$ in Eq. (1.88).

$i, j = 1, \ldots, n$, that

$$
b_{ij} = -\frac{\partial(\partial V/\partial a_i)}{\partial a_j} + \frac{\partial(\partial V/\partial a_j)}{\partial a_i} = 0,
$$

$$
b_{n+i,j} = 0 + \frac{\partial(\partial V/\partial a_j)}{\partial q_i},
$$

$$
b_{i,n+j} = -\frac{\partial(\partial V/\partial a_i)}{\partial q_j} + 0,
$$

$$
b_{n+i,n+j} = 0 - 0,
$$

$$
b_{i,2n+1} = -\frac{\partial(\partial V/\partial a_i)}{\partial t} - 0,
$$

$$
b_{2n+1,j} = 0 + \frac{\partial(\partial V/\partial a_j)}{\partial t},
$$

$$
b_{n+i,2n+1} = 0 - 0,
$$

$$
b_{2n+1,n+j} = 0 - 0,
$$

$$
b_{2n+1,2n+1} = 0 - 0. \tag{1.132}
$$

The first Pfaff system is thus, with a $(2n+1) \times (2n+1)$ matrix,

$$
\begin{pmatrix} 0 & -\dfrac{\partial(\partial V/\partial a_i)}{\partial q_j} & -\dfrac{\partial(\partial V/\partial a_i)}{\partial t} \\[2ex] \dfrac{\partial(\partial V/\partial a_j)}{\partial q_i} & 0 & 0 \\[2ex] \dfrac{\partial(\partial V/\partial a_j)}{\partial t} & 0 & 0 \end{pmatrix} \cdot \begin{pmatrix} \mathrm{d}a_j \\ \mathrm{d}q_j \\ \mathrm{d}t \end{pmatrix} = \begin{pmatrix} 0 \\ 0 \\ 0 \end{pmatrix}, \tag{1.133}
$$

or component-wise, for $i = 1, \ldots, n$,

$$
\sum_{j=1}^{n} \frac{\partial}{\partial q_j}\left(\frac{\partial V}{\partial a_i}\right) \mathrm{d}q_j + \frac{\partial}{\partial t}\left(\frac{\partial V}{\partial a_i}\right) \mathrm{d}t = 0, \tag{1.134}
$$

$$
\sum_{j=1}^{n} \frac{\partial}{\partial q_i}\left(\frac{\partial V}{\partial a_j}\right) \mathrm{d}a_j = 0, \tag{1.135}
$$

$$
\sum_{j=1}^{n} \frac{\partial}{\partial t}\left(\frac{\partial V}{\partial a_j}\right) \mathrm{d}a_j = 0. \tag{1.136}
$$

Equation (1.135) is

$$
\sum_{j=1}^{n} \frac{\partial^2 V}{\partial q_i \partial a_j} \mathrm{d}a_j = 0. \tag{1.137}
$$

Since the determinant of the $n \times n$ matrix $\partial^2 V/(\partial q_i \, \partial a_j)$ will in general not vanish, we conclude that for $i = 1, \ldots, n$,

$$
\mathrm{d}a_i = 0. \tag{1.138}
$$

Similarly, in Eq. (1.136) the coefficients $\partial^2 V/(\partial t\,\partial a_j)$ are independent, implying again Eq. (1.138). Finally, using Eqs. (1.134) and (1.138), one obtains

$$d\left(\frac{\partial V}{\partial a_i}\right) = \sum_{j=1}^{n}\frac{\partial}{\partial a_j}\left(\frac{\partial V}{\partial a_i}\right) + \sum_{j=1}^{n}\frac{\partial}{\partial q_j}\left(\frac{\partial V}{\partial a_i}\right)dq_j + \frac{\partial}{\partial t}\left(\frac{\partial V}{\partial a_i}\right)dt = 0. \quad (1.139)$$

Thus the equations of motion are, for $i = 1,\ldots,n$,

$$d\left(\frac{\partial V}{\partial a_i}\right) = 0, \qquad\qquad da_i = 0. \qquad\qquad (1.140)$$

That the a_i are n constants of motion is stated in the theorem, thus, contains no new information. Hence, there are n equations

$$\frac{\partial V}{\partial a_i} = b_i \qquad\qquad (1.141)$$

with n new constants of motion b_i. These equations can be solved by direct integration.   □

This proof of Liouville's theorem is somewhat unintuitive. We therefore consider the basic idea used in it, that each constant of motion reduces the number of Hamilton equations by *two*. This is first shown for the case that the constant of motion is energy, following Arnold (1989), pp. 242-243.

Lemma 7 *Let $H(p_1,\ldots,p_n,q_1,\ldots,q_n) = E$ be a constant of motion of a Hamiltonian system. Then only $2n - 2$ Hamilton equations need to be solved.*

Proof We assume that the equation $H = E$ can be solved for p_n (else use another of the p_i and rename it to p_n),

$$p_n = K(p_1,\ldots,p_{n-1},q_1,\ldots,q_{n-1},\tau;E), \qquad\qquad (1.142)$$

where

$$\tau = -q_n \qquad\qquad (1.143)$$

is a new time variable. Then for the Hamiltonian 1-form,

$$p_1\,dq_1 + \cdots + p_n\,dq_n - H\,dt =$$
$$p_1\,dq_1 + \cdots + p_{n-1}\,dq_{n-1} - K\,d\tau - d(Ht) + t\,dH. \qquad (1.144)$$

This equation shall define a canonical transformation from old variables on the left side to new variables on the right side. The total differential $d(Ht)$ can be dropped, as it does not enter the Hamilton equations. Furthermore, $dH = 0$ by assumption, and the last term can also be dropped. Thus the right side refers to canonical variables $p_1,\ldots,p_{n-1},q_1,\ldots,q_{n-1}$, a Hamilton function K and time τ. Demanding that the first Pfaff system of the Hamiltonian 1-form $p_1\,dq_1 + \cdots + p_{n-1}\,dq_{n-1} - K\,d\tau$ vanishes gives, in complete analogy with Eqs. (1.95) to (1.105) (with n replaced by $n-1$), the $2n-2$ Hamilton equations,

$$\frac{dq_i}{dq_n} = -\frac{\partial \tilde{K}}{\partial p_i}, \qquad \frac{dp_i}{dq_n} = \frac{\partial \tilde{K}}{\partial q_i} \qquad (i=1,\ldots,n-1) \qquad (1.145)$$

where

$$\tilde{K}(p_1,\ldots,p_{n-1},q_1,\ldots,q_{n-1};\tau,E) = K(p_1,\ldots,p_{n-1},q_1,\ldots,q_{n-1},\tau;E), \qquad (1.146)$$

meaning that $\tau = -q_n$ enters only as fixed a number in $\tilde{K}$, not as a variable. □

This lemma can be generalized to a single, arbitrary constant of motion. The following proof is taken from Nordheim & Fues (1927), pp. 115-116.

Lemma 8 *Let* $I(p_1,\ldots,p_n,q_1,\ldots,q_n) = a$ *be an arbitrary constant of motion of a Hamiltonian system. Then only $2n-2$ Hamilton equations need to be solved.*

Proof The idea is to find a canonical transformation to new coordinates P_i and Q_i and a new Hamiltonian K, so that the n-th momentum is constant, $P_n = a$. Then

$$0 = \frac{dP_n}{dt} = -\frac{\partial K}{\partial Q_n}, \qquad (1.147)$$

i.e., K does not depend on Q_n, and depends on P_n only as a constant parameter: the mechanical problem is thus reduced to $2n-2$ variables $P_1,\ldots,P_{n-1},Q_1,\ldots,Q_{n-1}$ and their corresponding Hamilton equations. To obtain this canonical transformation, assume an arbitrary function

$$F(Q_1,\ldots,Q_{n-1},q_1,\ldots,q_n), \qquad (1.148)$$

where F may also depend explicitly on t, and apply the canonical transformation from Eq. (1.19),

$$p_i = \frac{\partial F}{\partial q_i}, \qquad\qquad (i=1,\ldots,n),$$

$$P_j = -\frac{\partial F}{\partial Q_j}, \qquad\qquad (j=1,\ldots,n-1), \qquad (1.149)$$

which expresses the p_i as functions of the q_i and Q_j. These functions are inserted in the integral of motion I, to give a new function

$$J(q_1,\ldots,q_n,Q_1,\ldots,Q_{n-1}) = I(p_1,\ldots,p_n,q_1,\ldots,q_n) = a. \qquad (1.150)$$

Then

$$P_n = J \qquad (1.151)$$

accounts for the missing P_n in Eq. (1.149). The new Hamiltonian

$$K(P_1,\ldots,P_{n-1},Q_1,\ldots,Q_{n-1};a) = H(p_1,\ldots,p_n,q_1,\ldots,q_n) \qquad (1.152)$$

does not contain Q_n, and $P_n = a$ only as constant parameter.   □

The following theorem is a substantial generalization of Liouville's theorem, and shows that for each set of values of n independent integrals of motion $I_1, \ldots, I_n$ in involution, the trajectory lies on a torus in phase space.

Theorem (Arnold) *'Suppose that a Hamiltonian system with n degrees of freedom has n single-valued integrals of motion $H = I_1, I_2, \ldots, I_n$, standing pairwise in involution with one another. Suppose that the equations $I_i = a_i$ with $i = 1, \ldots, n$ distinguish in the 2n-dimensional space n-dimensional compact manifolds $M = M_a$, at each point of which the vectors grad I_i with $i = 1, \ldots, n$ are linearly independent. Then M is an n-dimensional torus and the point x depicting the solution of the Hamilton equations moves along it in a conditionally periodic manner.'* (Arnold 1963c, pp. 292-293, with slight alterations)

For the proof, see Arnold (1963c), Arnold (1989), pp. 271-278, or Lowenstein (2012), pp. 64-69. A weaker form of the theorem is proved in Katok & Hasselblatt (1995), pp. 227-228. The theorem is also treated in Abraham & Marsden (1978), pp. 393-396, who state (p. 396; they split the theorem into two):

> The two theorems of Arnold show us that complete integrability of a Hamiltonian system imposes serious restrictions on the topology of the manifold T^*M [in which trajectories lie].

The theorem is proved in two steps. In the first, it is shown (see sections 8.3.3 and 10.1.2 in Arnold 1989) that via the symplectic structure of phase space, the n linearly independent one-forms $\mathrm{d}I_i$ induce n vector fields $J \, \mathrm{d}I_i$, where J is the matrix introduced in Eq. (1.38). If the I_i are in involution, then the Hamiltonian phase flows along any two of these vector fields *commute*. In the second part of the proof, it is shown that a compact differentiable n-dimensional manifold M with n pairwise commutative and linearly independent vector fields is diffeomorphic to an n-dimensional torus.

As a simple example of the latter statement, consider a two-dimensional torus. Two circular, orthogonal cuts of the torus plus some stretching give a rectangle in the Euclidean plane, with commuting vector fields $\vec{e}_x$ and $\vec{e}_y$. For the two-dimensional sphere, nothing similar is possible, since the 2-sphere admits no continuous vector field of everywhere non-zero tangent vectors ('hairy ball theorem,' see Milnor 1978).

The occurence of a torus is, after all, not too surprising. In the older literature, the term 'multiple periodic,' as used in the following quotation, refers implicitly to a torus.

> Besonders einfach wird die Bewegung, wenn mindestens f [here: n] eindeutige Integrale existieren. [...] Mechanische Systeme dieser Art gehören, wenn man von Sonderfällen absieht und alle ins Unendliche verlaufenden Bewegungen

außer acht läßt, wohl immer zu den mehrfach periodischen.[12] (Fues 1927, p. 135)

The idea was obtained mainly from celestial mechanics, where motion is indeed often 'multiple periodic.'

We show in the following that integrals of motion in involution lead to commuting vector fields on a manifold. This will be of some help in Lemmas C2 and C3, which deal with Hamiltonian vector fields.[13]

Let M be a smooth manifold[14] with points x. Let $x(t) : (a,b) \subset \mathbb{R} \to M$ with $a < 0$ and $b > 0$ and $x_0 = x(0)$ be a differentiable curve in M.

Definition A *vector*[15] X_{x_0} at x_0 is the tangent to a curve through x_0,

$$X_{x_0} = \left. \frac{\mathrm{d}}{\mathrm{d}t} \right|_{t=0} x(t). \tag{1.153}$$

This equation is understood in a local coordinate system, given by a one-to-one mapping $U \to M$ (if M has dimension n) from an open set $U \subset \mathbb{R}^n$ to an open set in M including x_0. Then $x(t)$ and X are given by n-tuples of real or complex numbers, and $\mathrm{d}/\mathrm{d}t$ is well-defined.

The vector X_{x_0} belongs to a linear space of dimension n, the *tangent space* $T_{x_0}M$ 'attached' to M in x_0. A *vector field* is a map $M \to TM$ into the *tangent bundle* $TM = \bigcup_x T_x M$, associating with every point $x \in M$ a vector X_x in a differentiable fashion. On a *symplectic* manifold, with each function $F : M \to \mathbb{R}$ is associated a unique vector field X, as is shown below.

The most important concept in the following is the action of a vector field on differentiable real functions[16] $F, G, H : M \to \mathbb{R}$, the *Lie derivative*.

Definition Let $x(t) \in M$ be a smooth curve with $x_0 = x(0)$ and tangent vector X_{x_0} in x_0. Let $F : M \to \mathbb{R}$ be an arbitrary differentiable function. Then $X_{x_0}F$ is the derivative of F in the point x_0 in the direction of X,

$$X_{x_0}F = \left. \frac{\mathrm{d}}{\mathrm{d}t} \right|_{t=0} F(x(t)). \tag{1.154}$$

---

[12] 'The motion becomes particularly simple if at least f unique integrals [of motion] exist. [...] Mechanical systems of this type belong, refraining from special cases and leaving aside all motions to infinity, probably always to the multiple periodic.'

[13] Lemmas A1 to A4, G1 to G5, C1 to C4 and D1 to D7 are treated in Chapters 5 to 8.

[14] We skip over some standard definitions. Excellent introductions to differential geometry are Kobayashi & Nomizu (1963), Spivak (1970) (for the present context, especially chapter 5) and Sternberg (1964). The latter book has a section on Hamiltonian structures. Differential geometry for symplectic manifolds is treated in detail in Arnold (1989).

[15] To avoid excessive bracketing, we follow standard usage and write X_x (or, if no confusion can arise, X) instead of $X(x)$.

[16] The Hamilton function H indeed belongs here, one reason for this choice of function names.

The derivative on the right side is defined by standard calculus, since $F \circ x$ is a function $\mathbb{R} \to \mathbb{R}$. Equation (1.154) is (also) the definition of the *Lie derivative* $L_X F$ of F with respect to X.[17]

The derivative $X_x F$ is itself a function $M \to \mathbb{R}$, thus $Y(XF) = (Y \circ X)F$ is well-defined, and usually written YXF.

Definition Instead of just one curve $x(t)$ with tangent vectors X_x, one considers the *flow* $\phi_X^t(x_0) : \mathbb{R} \times M \to M$, the family of all curves with tangent vector X_x in each point $x \in M$. The family parameter is x_0, i.e., any point lying on a given curve (thus there are many references to a curve). The ϕ^t (we drop X if no confusion can occur) shall form a one-parameter *group* of diffeomorphisms with respect to the curve variable t,

$$\phi^0(x) = x,$$
$$\phi^{s+t}(x) = \phi^s(\phi^t(x)),$$
$$(\phi^t)^{-1}(x) = \phi^{-t}(x). \tag{1.155}$$

One says that the vector field X *generates* the flow ϕ_X^t. The flow ϕ_X^t generated by X is the family of *integral curves* of the vector field X. Equation (1.153) becomes then

$$X_{x_0} = \left.\frac{\mathrm{d}}{\mathrm{d}t}\right|_{t=0} \phi_X^t(x_0). \tag{1.156}$$

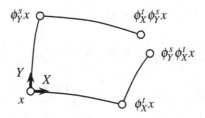

Figure 1.1 Two non-commutative vector fields

Two different flows on a manifold will in general not commute with each other (Fig. 1.1): moving from a point x with the flow generated by a vector field X over an

---

[17]For the identification $L_X F = XF$, see Dieudonné (1976), p. 304; Giaquinta & Hildebrandt (2004), p. 302; Kobayashi & Nomizu (1963), p. 29; Lang (1972), p. 110; Spivak (1970), p. 5-24; and Sternberg (1964), p. 92.

interval t and then with the flow generated by Y over an interval s, one will end up in some point y, but following first Y over s and then X over t, one will in general end up in some $z \neq y$. The following theorem gives a condition for two flows to commute. This is a fundamental theorem of differential geometry, and proved in Kobayashi & Nomizu (1963), p. 16; Lang (1972), pp. 107-108 (in parts); Spivak (1970), pp. 5-35, 5-36; and Giaquinta & Hildebrandt (2004), pp. 299-300. We follow here the proof given by Arnold (1989), pp. 210-213.

Theorem   *Let the vector fields X and Y generate the flows ϕ_X^t and ϕ_Y^s, respectively. Then*

$$\phi_X^t \circ \phi_Y^s = \phi_Y^s \circ \phi_X^t \quad \leftrightarrow \quad Y \circ X = X \circ Y. \tag{1.157}$$

Proof '$\rightarrow$'   Let $x_i : M \rightarrow \mathbb{R}$ with $x \mapsto x_i$ be the i-th coordinate function[18] on M. We perform a Taylor series expansion of the coordinate difference along the two alternative paths. Since this difference is zero if either $s = 0$ or $t = 0$, the first term in the expansion is of second order,

$$x_i(\phi_X^t(\phi_Y^s x)) - x_i(\phi_Y^s(\phi_X^t x)) =$$

$$st \left.\frac{\partial^2}{\partial s\,\partial t}\right|_{s=t=0} \left[x_i(\phi_X^t(\phi_Y^s x)) - x_i(\phi_Y^s(\phi_X^t x)) \right] \tag{1.158}$$

$$+ \frac{s^2 t}{2} \left.\frac{\partial^3}{\partial s^2 \partial t}\right|_{s=t=0} [\dots] + \frac{st^2}{2} \left.\frac{\partial^3}{\partial s \partial t^2}\right|_{s=t=0} [\dots] + \dots,$$

where the last set of dots stands for terms of order $s^2 t^2$ and higher. We calculate the second-order term, replacing $x(t)$ in Eq. (1.154) by either ϕ_X^t or ϕ_Y^s, and introducing auxiliary functions $F = Xx_i$ and $G = Yx_i$ from M to $\mathbb{R}$:

$$\left.\frac{\partial^2}{\partial s\,\partial t}\right|_{s=t=0} \left[x_i(\phi_X^t(\phi_Y^s x)) - x_i(\phi_Y^s(\phi_X^t x)) \right]$$

$$= \left.\frac{\partial}{\partial s}\right|_{s=0} \left.\frac{\partial}{\partial t}\right|_{t=0} x_i(\phi_X^t(\phi_Y^s x)) - \left.\frac{\partial}{\partial t}\right|_{t=0} \left.\frac{\partial}{\partial s}\right|_{s=0} x_i(\phi_Y^s(\phi_X^t x))$$

$$= \left.\frac{\partial}{\partial s}\right|_{s=0} Xx_i(\phi_Y^s x) - \left.\frac{\partial}{\partial t}\right|_{t=0} Yx_i(\phi_X^t x)$$

$$= \left.\frac{\partial}{\partial s}\right|_{s=0} F(\phi_Y^s x) - \left.\frac{\partial}{\partial t}\right|_{t=0} G(\phi_X^t x)$$

$$= YF(x) - XG(x)$$

$$= Y(Xx_i(x)) - X(Yx_i(x)). \tag{1.159}$$

Since terms with different powers of s and t will in general not cancel in Eq. (1.158), we can conclude: if $\phi_X^t \circ \phi_Y^s = \phi_Y^s \circ \phi_X^t$, then the second order term in Eqs. (1.158) and (1.159) must vanish, which gives $Y \circ X = X \circ Y$ by Eq. (1.159).

---

[18]The distinction between a point $x \in M$ and its coordinate $x_i \in \mathbb{R}$ should be clear from the given context.

'$\leftarrow$' The following beautiful argument is taken from Arnold (1989), pp. 212-213. For given finite values of s and t, the rectangle with corners $(0,0)$, $(s,0)$, $(0,t)$, (s,t) in the st-plane is subdivided into N^2 small rectangles (Fig. 1.2). Changing path 1 from point a to point b (first s, then t) into path 2 (first t, then s) 'continuously,' requires to switch from each traversed side to the opposing side in all of the N^2 small rectangles, as is indicated in the figure. From Eqs. (1.158) and (1.159), the difference in x_i along the two paths from one corner to the corner diagonally opposite in a small rectangle includes terms with $s^2 t/N^3$ and st^2/N^3 (and N^{-4}, etc.), if $Y \circ X = X \circ Y$. Thus the difference in x_i between paths 1 and 2 from a to b scales as $N^2 \times N^{-3} = N^{-1}$, and vanishes for $N \to \infty$ if $Y \circ X = X \circ Y$. $\square$

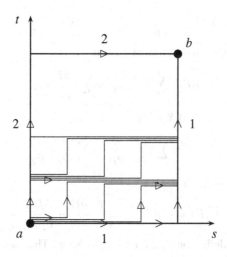

Figure 1.2 A continuous deformation of path 1 into path 2 requires alterations to opposing sides in N^2 small squares, from arrows $>$ to $\triangleright$ (tiny distances in the figure are meant to vanish)

We turn now to mechanical systems. A *symplectic* manifold, most notably phase space, is a manifold with *even* dimension and with an isomorphism from differential one-forms to vector fields given by

$$dF \mapsto X = J\, dF, \qquad (1.160)$$

where J is the matrix from Eq. (1.38). The vector field $J\, dF$ is called a *phase velocity*, the most important phase velocity is that associated with the Hamilton function, $J\, dH$, and all flows on symplectic manifolds are termed *Hamiltonian phase flows*. Equations (1.154) and (1.156) remain valid with $X = J\, dF$. The connection to classical mechanics is via the relation (see Arnold 1989, p. 214)

$$\{G,F\} = XG \qquad \text{with} \qquad X = J\, dF. \qquad (1.161)$$

Here $\{.,.\}$ is the *Poisson bracket* of two functions in phase space. To prove Eq. (1.161), assume that the symplectic manifold M is $\mathbb{R}^{2n}$ with Cartesian coordinates (p_i, q_i) where $i = 1, \ldots, n$. Then the canonical basis of the tangent space is (see Kobayashi & Nomizu 1963, p. 4)

$$\left(\frac{\partial}{\partial p_1}, \ldots, \frac{\partial}{\partial p_n}, \frac{\partial}{\partial q_1}, \ldots, \frac{\partial}{\partial q_n} \right), \tag{1.162}$$

and the canonical basis of the dual space of 1-forms is

$$(\mathrm{d}p_1, \ldots, \mathrm{d}p_n, \mathrm{d}q_1, \ldots, \mathrm{d}q_n). \tag{1.163}$$

The isomorphism from 1-forms to vectors is given by

$$\mathrm{d}p_i \mapsto -\frac{\partial}{\partial q_i}, \qquad \mathrm{d}q_i \mapsto \frac{\partial}{\partial p_i}. \tag{1.164}$$

For a function $F(p,q)$ with differential 1-form

$$\mathrm{d}F = \sum_{i=1}^{n} \frac{\partial F}{\partial p_i} \mathrm{d}p_i + \sum_{i=1}^{n} \frac{\partial F}{\partial q_i} \mathrm{d}q_i \tag{1.165}$$

the corresponding vector field X is thus

$$J\,\mathrm{d}F = \sum_{i=1}^{n} \frac{\partial F}{\partial q_i} \frac{\partial}{\partial p_i} - \sum_{i=1}^{n} \frac{\partial F}{\partial p_i} \frac{\partial}{\partial q_i}. \tag{1.166}$$

Equation (1.161) becomes then

$$\{G,F\} = (J\,\mathrm{d}F)G = \frac{\partial F}{\partial q_i} \frac{\partial G}{\partial p_i} - \frac{\partial F}{\partial p_i} \frac{\partial G}{\partial q_i}, \tag{1.167}$$

which is indeed the definition of the Poisson bracket. The above theorem for commuting flows takes now the following form.

Theorem *Let vector fields $X = J\,\mathrm{d}F$ and $Y = J\,\mathrm{d}G$ generate Hamiltonian phase flows ϕ_X^t and ϕ_Y^s on a symplectic manifold. Then*

$$\phi_X^t \circ \phi_Y^s = \phi_Y^s \circ \phi_X^t \quad \leftrightarrow \quad \{G,F\} = 0. \tag{1.168}$$

Proof Equation (1.159) is

$$\frac{\partial^2}{\partial s\,\partial t}\bigg|_{s=t=0} \left[x_i(\phi_X^t(\phi_Y^s x)) - x_i(\phi_Y^s(\phi_X^t x)) \right] = Y\big(Xx_i(x)\big) - X\big(Yx_i(x)\big).$$

We evaluate the right side further using Eq. (1.161),

$$\begin{aligned}
Y\big(Xx_i(x)\big) - X\big(Yx_i(x)\big) &= Y\big((J\,\mathrm{d}F)x_i(x)\big) - X\big((J\,\mathrm{d}G)x_i(x)\big) \\
&= Y\{x_i, F\}(x) - X\{x_i, G\}(x) \\
&= (J\,\mathrm{d}G)\{x_i, F\}(x) - (J\,\mathrm{d}F)\{x_i, G\}(x) \\
&= \{\{x_i, F\}, G\}(x) - \{\{x_i, G\}, F\}(x) \\
&= \{\{G, F\}, x_i\}(x). \tag{1.169}
\end{aligned}$$

In the last line, $\{F,G\} = -\{G,F\}$, and the Jacobi identity

$$\{\{F,G\},H\} + \{\{G,H\},F\} + \{\{H,F\},G\} = 0 \qquad (1.170)$$

were used (not proved here). With Eq. (1.169), the present theorem follows from the last theorem. $\qquad\qquad\qquad\qquad\qquad\qquad\qquad\qquad\qquad\qquad\qquad\qquad\qquad\Box$

1.8 KNESER'S THEOREMS FOR FOUR-DIMENSIONAL PHASE SPACE

In four-dimensional phase space, i.e., for two degrees of freedom, the restriction of trajectories to two-dimensional tori was proved by Kneser (1921, 1924) using Euler's polyhedron formula.[19]

Definition A family of curves in a two-dimensional surface F is *regular*, if for every neighborhood $U \subset F$ there is a homeomorphism that maps U to a neighborhood U' in the Euclidean plane and the curves segments in U to parallel straight line segments in U'.

Theorem (Kneser) *A closed, two-dimensional surface with an everywhere regular family of curves is either a torus or a Klein bottle.*

Proof A line net is established on the surface according to the following rules:

1. A finite number q of line segments called *bars* is put across the regular curves. Regular curves and bars are nowhere tangent. The distance between two bars is below some maximum.
2. The regular curves through each of the two endpoints of a bar are followed in each direction until they meet the next bar.
3. This 'next' juncture of a curve with a bar is not at one of the two bar endpoints (else change the bar length).
4. No two curve segments end in the same point on a bar from different directions (else change a bar length).

The bar endpoints and the endpoints of curve segments on bars define *corners*. Between two corners lies one *edge*, and ≥ 4 edges enclose an *area*. Figure 1.3 gives an example of this construction *assuming* a torus or a Klein bottle. The regular curves are here straight vertical lines, the bars straight horizontal lines. The two vertical boundary lines in the figure are to be identified, as are the two horizontal boundary lines; this leads to a torus. Identifying the two horizontal boundaries 'crosswise' gives a Klein bottle.

---

[19]Kneser (1921) is somewhat overlooked, probably because it deals with Bohr-Sommerfeld quantization, but is cited in Fues (1927), p. 135. Kneser (1924) is cited in Kolmogorov (1957), Abraham & Marsden (1978) and in the literature on dynamical systems on phase-space surfaces, e.g., Siegel (1945), Godbillon (1983) and Hector & Hirsch (1986).

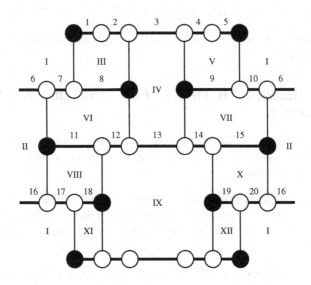

Figure 1.3 Line net on a torus or Klein bottle. The horizontal and vertical boundaries are to be identified. The Arabic numerals count the horizontal edges on bars (fat lines), the Roman numerals the areas

The number of corners is $c = 6q$: there are $2q$ endpoints of bars, and from each of these endpoints, two curve segments start, ending on the next bar, giving another $4q$ corners.

The number of edges is $e = 9q$: there are q bars, and from their endpoints, a total of $4q$ curve segments start. Each of them ends on a bar. In this junction, the curve segment increases the number of bar segments by one, giving another $4q$ segments.

The number of areas is $a = 3q$: each bar endpoint (full circle) belongs to three different areas; and to each area belong two bar endpoints, one on each of its two vertical edges. Thus $2q$ bar endpoints belong to $6q$ areas, including multiple counting. With each area containing two bar endpoints, there are thus $3q$ different areas. Euler's polyhedron formula is

$$c - e + a = 2 - 2g, \qquad (1.171)$$

where g is the *genus* of the polyhedron. The left side is $6q - 9q + 3q = 0$, thus $g = 1$ ('one tunnel'), which means the closed surface is a torus or Klein's bottle.   □

Remark For comparison, in the torus of Fig. 1.3 there are 8 different full and 16 different open circles (i.e., not counting the lowest row), thus $c = 24$ corners. There are 20 horizontal and 16 vertical edges (4 in each row) between corners, giving $e = 36$. And there are $a = 12$ areas (Roman numerals). Thus indeed

$$c - e + a = 24 - 36 + 12 = 0. \tag{1.172}$$

The next theorem (Kneser 1921) will not be used in the rest of the book. We include it for the originality of the proof.

Theorem (Kneser) *Given is a Hamiltonian system with two degrees of freedom and two independent integrals of motion $\Phi(q_1, q_2, p_1, p_2)$ and $\Psi(q_1, q_2, p_1, p_2)$. There shall be a one-to-one correspondence between pairs (a, b) in an open neighborhood of $\mathbb{R}^2$ and a family of nested, two-dimensional tori $F_{ab} = \{(q_1, q_2, p_1, p_2) : \Phi = a, \Psi = b\}$. The trajectories in each F_{ab} shall be regular, and each F_{ab} contains a closed trajectory. Then each F_{ab} contains a ring in which all trajectories are closed.*

Remark This ring is called *Kneser ring*, see Anosov (1995), p. 218.

Proof (Kneser 1921, pp. 296-297) Let U be a four-dimensional neighborhood in phase space of a point P on a closed trajectory C in one of the F_{ab}. Introduce local coordinates Φ, ψ, x, y in U, all of them analytic functions of q_1, q_2, p_1, p_2. Let y be constant on segments of trajectories lying in U, especially $y = 0$ for $C \cap U$. By narrowing U on F_{ab}, one has $y = 0$ also for the next re-entry of C into U, since a closed curve cannot be dense (the re-entry is typically at an x-value different from that of P.) Following other (open, winding) trajectories in $F_{ab} \cap U$, they will generally have different y upon re-entry in U. For the k-th re-entry, this value shall be y_k. Then $y_k(y, \Phi, \Psi) - y$ is a differentiable function, and its zeros correspond to closed trajectories.

For each k, the equation $y_k - y = 0$ defines a two-dimensional surface Y in the three-dimensional space of variables y, Φ, Ψ,

$$y = y_k(y, \Phi, \Psi) \quad \leftrightarrow \quad y = Y(\Phi, \Psi) \tag{1.173}$$

with differentiable Y. Some of the Y may degenerate to line segments or points; but there must be some k for which Y is a two-dimensional surface with independent variables Φ and Ψ, since $(a, b) \mapsto F_{ab}$ is one-to-one, and each F_{ab} has a closed trajectory.

Choose a specific $y = Y + \varepsilon$ in U. It can be assumed that $y_k > Y$, see Fig. 1.4. In the left panel, the function $(y_k - y)(y)$ has slope zero at $y = Y$ (Φ and Ψ are constant parameters). Thus $y_k(y) = (y_k - y) + y$ has slope one at $y = Y$, and $y > Y$ implies $y_k > Y$ in a neighborhood of Y. The dotted line with $y > Y$ and $y_k < Y$ cannot occur since $y_k - y$ is not differentiable at Y. It would also imply that two different y have the same y_k, thus, trajectories merge, violating uniqueness. In the right panel, one has $y > Y$ but $y_k < Y$. Choose $y' = y_k$ as new starting point. Then $y'_k = y_{2k} > Y$, and consider y and y_{2k} in the following argument.

The function $y_k - y$ is differentiable, does not vanish identically and has a zero at $y = Y(\Phi, \Psi)$. There exists then a three-dimensional domain E (after possible further

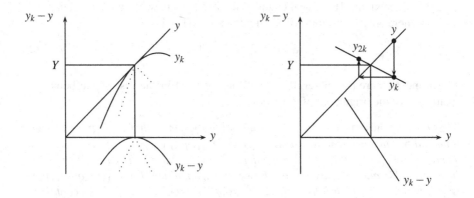

Figure 1.4 The function $y_k(y, \Phi, \Psi)$ at fixed Φ, Ψ

reduction of ε),

$$Y < y \le Y + \varepsilon, \qquad |\Phi - a| \le \varepsilon, \qquad |\Psi - b| \le \varepsilon, \qquad x = 0, \qquad (1.174)$$

in which $y_k - y$ has the same sign for all y. Assume that $y_k < y$ (altogether $Y < y_k < y \le Y + \varepsilon$); this will lead in the following to the contradiction that the phase-space measure decreases during motion (cf. Section 1.4). If instead $y_k > y$ for all y in the considered domain, this will lead to the contradiction that the phase-space measure increases during motion.

For each point in the domain E, follow the trajectory through this point to its k-th re-entry into U at some $x \ne 0$ and $y_k < y$, and follow it further until $x = 0$ (with y_k staying constant). This defines a subset $E' \subset E$ of smaller measure than E (since $y_k < y$). The set of all trajectory points defines a four-dimensional domain D in phase space, starting on E and ending on E'. (D resembles a contracting tube, see Fig. 1.5.)

Assume now that this set D is a four-dimensional domain of initial conditions in phase space. Following each point of D for a fixed time $\tau > 0$, one obtains a domain D_τ with smaller measure than D. Indeed, D contains by construction the future positions of all of its points, especially those with $x \ne 0$ that will reach E' within τ. This decrease of phase-space measure contradicts the theorem on conserved phase-space measure, and one concludes that all trajectories through E must be closed.   □

1.9 CELESTIAL MECHANICS

Much of the work that led to the KAM theorem was motivated by the question, whether the solar system is stable. The theory of orbits of gravitating bodies in the solar system is a central topic of celestial mechanics. Over the last decades, asteroids

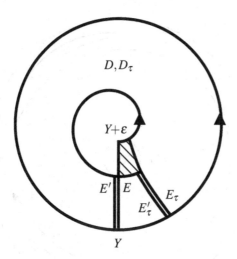

Figure 1.5 Decrease of phase-space measure. The four-dimensional domain D extends from the three-dimensional domain E to its true subset $E' \subset E$ (drawn with small offset; the outermost points of E and E' are connected by two curves with arrows). The domain D_τ extends from E_τ to its subset E'_τ. Thus D_τ is D minus the hatched region

up to a few kilometers in diameter were discovered that orbit the Sun in near 1:1 resonance with Earth on so-called horseshoe orbits. Figure 1.6 shows the simplest form of a horseshoe in the Roche potential of a binary system given by

$$V(r) = -\frac{Gm_1}{r_1} - \frac{Gm_2}{r_2} - \frac{\omega^2 r^2}{2}, \qquad (1.175)$$

with gravitational constant G, masses m_1 and m_2 of the gravitating bodies in circular orbits about their common center of mass, r the distance of a point from this center of mass and with r_1 and r_2 the distances respectively from m_1 and m_2 (with $m_1/m_2 = 100$ in the figure). The last term in Eq. (1.175) is the centrifugal potential in a frame that rotates with the angular frequency ω of the line connecting m_1 and m_2. The positions of m_1 and m_2 in the figure are thus fixed.

The small, upper and lower elongated lobes in Fig. 1.6 correspond to *tadpole* orbits of satellites that move around one of the stable Lagrange points marked '+'. The masses m_1 and m_2 and each of the two Lagrange points form an equilateral triangle. The *horseshoe* orbit is the double circle around m_1 that avoids the region close to m_2. It results when two extremely elongated tadpoles open up and connect with each other. It winds around m_1 enclosing both Lagrange points.

The period of a satellite (i.e., an asteroid or moon) on a horseshoe orbit is close to the period T of m_2 (since the distances from m_1 are comparable). On the inner circle,

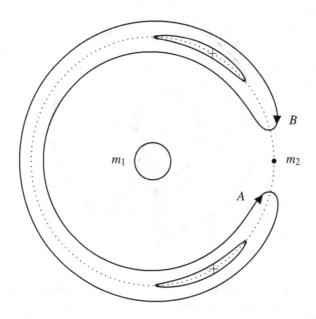

Figure 1.6 Horseshoe and tadpole orbits

the satellite period is $< T$, and it is $> T$ on the outer circle. Assuming $\omega > 0$, in point A of the figure the satellite 'drifts' (librates) slowly along the horseshoe toward m_2. Since the Coriolis acceleration proportional to the velocity relative to the rotating frame is not included in Eq. (1.175), the drift on the horseshoe must be very slow to make this so-called zero-velocity curve a good approximation to the real orbit. The gravitational pull of m_2 accelerates the satellite and transfers it to a higher orbit, i.e., the outer circle (see next paragraph). Here the satellite lags behind the rotation of m_1 and m_2 and drifts toward B. In B, the satellite moves still, as always, in the positive sense (counter-clockwise) on its orbit. Thus the gravitational pull from m_2 decelerates the satellite now, and transfers it to the inner circle on which it drifts toward A. A complete transfer around the horseshoe may take a few hundred periods T.

The transfer between two circular orbits by two accelerations in flight direction is known as a *Hohmann transfer*, see Fig 1.7. The first acceleration a_1 brings the satellite S from the inner circle to the Kepler ellipse E (tangent to the circle in the acceleration point). Having moved halfway around the ellipse, a second acceleration a_2 brings the satellite on the outer circular orbit. Since the azimuthal speed on a circular orbit is $v_\phi \sim r^{-1/2}$, the satellite is slower on the outer orbit. Counter-intuitively, thus, two accelerations lead to slower speed. The kinetic energy of the satellite drops

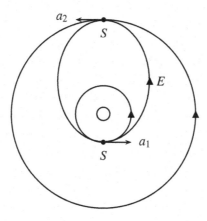

Figure 1.7 Hohmann transfer between circular orbits

by $\frac{1}{2}GM(R^{-1} - r^{-1}) < 0$ during the transfer (inner and outer orbit radius r and R, respectively, and mass M of the central body), but the gravitational binding weakens by twice this amount, $GM(r^{-1} - R^{-1}) > 0$, thus the total energy of the satellite increases.

In the solar system, a number of asteroids and moons are known to be on horseshoe orbits. The largest object in resonance with Earth is 3753 Cruithne, with a diameter of ≈ 5 km and at a distance between 0.48 and 1.51 AU from the Sun (Earth: 1 AU). Its minimum distance from Earth is ≈ 30 times that of the Moon, its period around the Sun is 364 days, and one passage through the horseshoe orbit takes 770 years (numbers from Wikipedia entry '3753 Cruithne,' which has also an animation of the drift of Cruithne's solar ellipse around the horseshoe).

The question of stability of horseshoe orbits is subject of current research. Horseshoes in systems with mass ratios $m_1/m_2 > 1200$ (Ćuk et al. 2012) can be long lived, e.g., exist for billions of years; whereas for smaller mass ratios, horseshoes are unstable because they approach the secondary m_2 too closely. Thus, horseshoes of bodies in 1:1 resonance with Saturn ($m_1/m_2 \approx 3500$ with respect to the Sun) should be long term stable, whereas those in resonance with Jupiter ($m_1/m_2 \approx 1000$) tend to be unstable (Laughlin & Chambers 2002).

As a near 1:1 resonance, the horseshoe orbit is not subject of the KAM theorem. While resonances are excluded in the KAM theorem by a Diophantine condition, one should still note that

there is a growing realization among planetary scientists that an improved understanding of these intimate 1:1 relationships is essential to studies ranging from planetary accretion to ring dynamics. (Murray 1997, p. 652)

2 Preliminaries

2.1 NOTATION

This section introduces the notations and conventions from section 4.5 in Arnold (1963a).

Functions All functions have complex arguments and complex values. If the argument is real, the function value is also real. All functions are assumed to be *complex analytic*, i.e., complex differentiable without poles or essential singularities.

The variables are the canonical variables $p = (p_1, \ldots, p_n)$ (generalized momenta) and $q = (q_1, \ldots, q_n)$ (generalized positions), all considered to be complex numbers, where $n \geq 2$ is the number of degrees of freedom of the dynamical system. These variables are abbreviated as

$$x = (x_1, \ldots, x_{2n}) = (p_1, \ldots, p_n, q_1, \ldots, q_n), \tag{2.1}$$

with momenta always written first.

The Hamiltonian $H^+(p)$ of the unperturbed system depends on the momenta p alone, not on the positions q. This means that the p are action variables and the q are angle variables. The range of each q_i is from 0 to 2π, and all functions have period 2π in each q_i *by assumption*. The n frequencies of the periodic motions of the integrable system are given by

$$\omega_i = \frac{\partial H^+(p)}{\partial p_i}. \tag{2.2}$$

Norms for vectors are introduced by[1]

$$|p| = \max_{i=1,\ldots,n} \{|p_i|\}, \qquad |q| = \max_{i=1,\ldots,n} \{|q_i|\}, \qquad |\omega| = \max_{i=1,\ldots,n} \{|\omega_i|\}. \tag{2.3}$$

This maximum norm is slightly unconventional (or inconvenient) in geometric considerations, but it is of advantage in analytic estimates, where one needs to consider only one component of vectors then, that of maximum magnitude. The maximum norm of a vector *induces* a natural matrix norm by

$$|D| = \max_{x \neq 0} \frac{|D \cdot x|}{|x|} = \max_{|x|=1} |D \cdot x|, \tag{2.4}$$

with $D = (d_{ij})$ an arbitrary $(m \times n)$-matrix, x an arbitrary n-vector and $\cdot$ the matrix product. This induced matrix norm is *compatible* with the vector norm in the sense

---

[1]To avoid elaborate notation, we write vector norms as $|p|$, as for the magnitude of a vector, and use $\|f\|$ for function norms only.

DOI: 10.1201/9781003287803-2

that, according to the first equality in Eq. (2.4),

$$|D \cdot x| \leq |D| \, |x| \tag{2.5}$$

for arbitrary D and x. Since $|(\pm 1, \ldots, \pm 1)| = 1$ for the maximum norm (each component is independently $+1$ or -1), the maximum in Eq. (2.4) is obtained for one of these vectors,

$$|D| = \max_{i=1,\ldots,m} \left(\sum_{j=1}^{n} |d_{ij}| \right), \tag{2.6}$$

which is called *row-sum norm*. From this follows, with n the number of columns,

$$|D| \leq n \max_{i,j} (|d_{ij}|), \tag{2.7}$$

an estimate used in the following.

Arnold (1963a) does not explicitly introduce a norm for functions, but Arnold (1963b, p. 172) does: '$A \leq E$ denotes that $|Ax| \leq |x|$ for any x.' Therefore, let

$$\|A\| = \sup_{x \in \text{dom } A} \{|A(x)|\}, \tag{2.8}$$

where 'dom A' is the domain of A. Compact domains are considered in the following, thus any function attains a maximum on its domain. In Eq. (2.8), $A(x)$ is a real or complex number a with magnitude $|a|$, or a complex vector for which Eq. (2.3) applies. Note that if $\|f_n - f\| \to 0$ in the norm (2.8), then $f_n \to f$ *uniformly*.

Because of the periodicity in q, functions can be expanded in a Fourier series with respect to the q-coordinate,

$$f(p,q) = \sum_{k} f_k(p) e^{i(k,q)}, \tag{2.9}$$

with scalar product $(k,q) = k_1 q_1 + \cdots + k_n q_n$, where $k = (k_1, \ldots, k_n)$ is a tuple in $\mathbb{Z}^n$ (including zero), and the sum is taken over all values $-\infty < k_i < \infty$. If $k = 0$ is taken outside the sum, one writes

$$f(p,q) = f_0(p) + \sum_{k}{}' f_k(p) e^{i(k,q)}, \tag{2.10}$$

where $\sum_{k}'$ is over all values of $k = (k_1, \ldots, k_n)$ not including $(0, \ldots, 0)$. In k-space we choose the 1-norm,[2]

$$|k| = |k_1| + \cdots + |k_n|. \tag{2.11}$$

Partial derivatives are abbreviated as

$$f_p = \frac{\partial f}{\partial p} = \left(\frac{\partial f}{\partial p_1}, \ldots, \frac{\partial f}{\partial p_n} \right). \tag{2.12}$$

---

[2]The 1-norm $|k|$ appears in Lemma A2, Eq. (5.24) and is used to specify the highest resonances that are excluded at a given iteration step. Note again that Lemmas A1 to A4, G1 to G5, C1 to C4 and D1 to D7 are treated in Chapters 5 to 8.

Domains All domains (say for $(p,q) \in \mathbb{C}^{2n}$) are connected and compact. The latter means that they are bounded and include all boundary points. These boundaries shall consist of a finite number of differentiable hypersurfaces (manifolds). *Complex analytic* functions f are defined on *open* connected domains:

> A function f, defined in a [open] domain D, is said to be *holomorphic* (analytic) at a point $z_0 \in D$ if there exists a neighbourhood of this point in which f may be represented by a power series. (Gonchar 1995, p. 169)

A function f is said to be holomorphic (analytic) in D if it is so in each $z \in D$. This is generalized to *closed* domains as follows:

> The function f is called analytic on the set E if it is analytic on some open set which contains E (or, more exactly, if there exist both an open set containing E and an analytic function F on this set which coincides with f on E). For open sets the notion to analyticity coincides with the notion of differentiability with respect to the set. However, this is not the case in general. (Gonchar 1995, p. 172)

Maps 'The mappings we consider are given by analytic functions'[3] (Arnold 1963a, p. 30). A *diffeomorphism* f is a one-to-one map where both f and f^{-1} are (continuously, if f is real) differentiable at each point of the domains of f and of f^{-1}, respectively. The following three diffeormorphisms appear frequently: (i) the map from momentum/action variables to frequencies,

$$A : G \to \Omega, \qquad p_i \mapsto \omega_i = \frac{\partial H^+(p)}{\partial p_i}, \tag{2.13}$$

(ii) the canonical transformation of coordinates (p,q),

$$B_s : F_s \to F_{s-1}, \qquad (p_s, q_s) \mapsto (p_{s-1}, q_{s-1}), \tag{2.14}$$

and (iii) the composition $S_s = B_1 \circ B_2 \circ \cdots \circ B_s$. The identity map is given the symbol E, $E(x) = x$.

Constants The real numbers $\theta, \Theta, \rho, \kappa, D$ are positive and of order unity. Their meaning is roughly as follows.

---

[3]We consider 'function' and 'mapping' to be equivalent terms (and usually abbreviate the latter as 'map'): 'Each set $f = \{(x,y)\}$ of ordered pairs $(x,y), x \in X, y \in Y$, such that ... is called a *function* or, what is the same thing, a *mapping*. As well as the terms 'function' and 'mapping' one uses in certain situations the terms 'transformation', 'morphism', 'correspondence', which are equivalent to them' (Kudryavtsev 1995, p. 694). Or: 'Logically, the concept of a 'mapping' coincides with the concept of a *function*, an *operator* or a *transformation*' (Sobolev 1995, p. 745). Or: 'A functional graph [i.e., $x \mapsto y$ is single-valued] in $X \times Y$ is also called a *mapping of X into Y*, or a *function defined in X, taking its values in Y*' (Dieudonné 1969, p. 5).

θ, Θ: lower and upper limit of the slope of the frequency diffeomorphism A from Eq. (2.13),

$$\theta \, |dp| \leq \|dA\| \leq \Theta \, |dp|, \tag{2.15}$$

with

$$0 < \theta < 1 < \Theta < \infty. \tag{2.16}$$

ρ: the half-width of the imaginary strip of the spatial coordinate q (note the intended similarity of the letters ρ and q). The Hamiltonian is assumed to be analytic in the domain

$$|\text{Im } q| \leq \rho. \tag{2.17}$$

$\kappa < 1$: gives the measure of a subset as fraction of the measure of the full set. Especially

$$\text{mes} \, (G \backslash G_s) \leq \kappa \, \text{mes} \, (G), \tag{2.18}$$

where G is the domain of action variables p and $G_s \subset G$ is the subdomain at iteration step s, for which the motion is on a phase-space torus. The number κ is freely chosen, and other small quantities depend on κ. D: the type of a domain (Arnold 1963a, sect. 4.1.1°). D has the unit of an inverse length.

The positive real numbers $\kappa, K, M, \delta, \beta, \gamma$ are control parameters for different limit processes and thus very small. The numbers ρ, θ, Θ are positive and not necessarily small. $N \in \mathbb{N}$ is very large, $N \gg 1$, and is the upper bound, $|k| < N$, for the vector $k \in \mathbb{Z}^n$. The numbers $\nu, \nu_1, \nu_2, L, L_0, \ldots, L_5, T$ are constants that depend only on n, e.g., $\nu = 2n + 3$ and $T = 8n + 24$. The iteration index for the infinite sequence of canonical transformations is $s \in \mathbb{N}$.

In lists we use the following 'numbering':

(A), (B), (C), etc., for assumptions in a lemma or theorem

(I), (II), (III), etc., for conclusions in a lemma or theorem

(i), (ii), (iii), etc., for steps in a proof and for general lists

(a), (b), (c), etc., for individual equations in equation sequences

The basic objects in the KAM theorem are the (iterated) domains F_s, G_s, Ω_s and functions A_s, B_s, H_s, S_s on them:

F_s: $2n$-dimensional domain of canonical variables (p_s, q_s)

G_s: n-dimensional subdomain of F_s of momenta p_s

Ω_s: frequency domain $\omega = \partial H_s^+ / \partial p_s$.

$A_s : G_s \to \Omega_s$: diffeomorphism from momentum to frequency, $p_s \mapsto \omega$

$B_s : F_s \to F_{s-1}$: canonical transformation $(p_s, q_s) \mapsto (p_{s-1}, q_{s-1})$

$H_s : F_s \to \mathbb{C}$: Hamiltonian after s-th canonical transformation

S_s : generating function of transformation $(p_s, q_{s-1}) \mapsto (p_{s-1}, q_s)$

2.2 DOMAIN REDUCTION

To obtain estimates for derivatives using the Cauchy theorem from complex function theory, points p must have a minimum distance, $\beta > 0$, from the boundary of a domain G; this is written as $G - \beta$. The exact definition is that 'p with a β-neighbourhood belongs to G' (Arnold 1963b, p. 105) or '$G - \beta$ denote[s] the set of points contained in G together with a β-neighbourhood'[4] (Arnold 1963a, p. 30; symbols adapted). Thus,

$$p \in G - \beta \quad \leftrightarrow \quad \overline{B}_\beta(p) \subset G, \tag{2.19}$$

where $\overline{B}_\beta(p)$ is the closed ball (a cube in the maximum norm) with radius β centered at p. The domain $G - \beta$ is obtained by excluding open balls $B_\beta(p)$ from G for every boundary point $p \in \partial G$. It is of some importance that $G - \beta$ is *closed*, i.e., that open balls are removed from G. The reason is that then, after all subset removals, a closed, nowhere dense, Cantor set remains in the statement of the KAM theorem on which, nevertheless, differentiability can be defined.

Lemma

$$G - \beta = G \setminus \bigcup_{q \in \partial G} B_\beta(q). \tag{2.20}$$

Proof '$\setminus$' is set subtraction. For $p \in G$, consider the two cases

$$
\begin{array}{ll}
p \in G - \beta & p \notin G - \beta \\
\xrightarrow{} \overline{B}_\beta(p) \subset G & \xrightarrow{} \overline{B}_\beta(p) \not\subset G \\
\overset{(\P)}{\xrightarrow{}} \forall q \in \partial G \quad |p - q| \geq \beta & \xrightarrow{} \exists q \in \partial G \quad |p - q| < \beta \\
\xrightarrow{} \forall q \in \partial G \quad p \notin B_\beta(q) & \xrightarrow{} p \in B_\beta(q) \\
\xrightarrow{} p \in G \setminus \bigcup_{q \in \partial G} B_\beta(q) & \xrightarrow{} p \notin G \setminus \bigcup_{q \in \partial G} B_\beta(q).
\end{array}
$$

To see that ($\P$) holds, assume there is a $q \in \partial G$ with $\eta = |p - q| < \beta$. By definition of the boundary of a domain, every (arbitrarily small) neighborhood of $q \in \partial G$ contains points $r \notin G$. Choose $\varepsilon > 0$ so small that $\eta + \varepsilon < \beta$ and select a point $r \notin G$ but $r \in B_\varepsilon(q)$, i.e., $|q - r| < \varepsilon$. Then

$$|p - r| = |p - q + q - r| \leq |p - q| + |q - r| < \eta + \varepsilon < \beta, \tag{2.21}$$

thus $r \notin G$, but $r \in B_\beta(p) \subset G$, i.e., $r \in G$; contradiction. □

---

[4]Or slightly more specific: $U - d$ is the set of points $p \in U$ so that p together with a d-neighborhood of p is included in U. Alternatively, with 'dist' the distance:

Let $\Gamma \subset \mathbb{R}^n$ be an open subset [...] We also consider the shrunken version of Γ defined by $\Gamma_\gamma = \{\omega \in \Gamma | \text{dist}(\omega, \partial \Gamma) \geq \gamma\}$. (Broer & Sevryuk 2010, p. 270)

Similarly, '$G+\beta$ denote[s] the β-neighbourhood of G' (Arnold 1963, p. 30; symbols adapted), i.e., the union of G and all closed (!) β-balls centered on points $p \in G$. Clearly, the latter G can be replaced by ∂G,

$$G+\beta = G \cup \bigcup_{q \in \partial G} \bar{B}_\beta(q). \tag{2.22}$$

Note that in general one only has

$$(G-\beta)+\beta \subset G, \tag{2.23}$$

but no set equality of the left and right side, as is shown in Fig. 2.1.

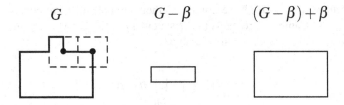

$$G \qquad\qquad G-\beta \qquad\qquad (G-\beta)+\beta$$

Figure 2.1 Demonstration of $(G-\beta)+\beta \subset G$. The two squares with dashed boundary lines are balls B_β (maximum norm) taken off from G (bold line) at the points marked with bullets

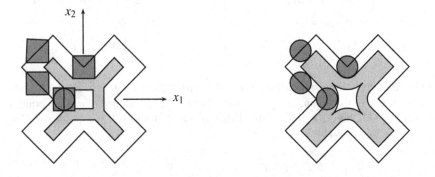

Figure 2.2 Taking off two layers of width β from a domain G (outer white region), resulting in a light gray and an inner white domain. *Left:* maximum norm; *right:* 2-norm. Squares and circles in dark gray are balls B_β in the respective norm

In Fig. 2.2, two layers of constant width β are taken off a cross-shaped domain using the sup norm of Eq. (2.3) and, for comparison, the Pythagorean 2-norm.

Let Re Ω be the intersection of the complex domain $\Omega \subset \mathbb{C}^n$ with the real space $\mathbb{R}^n$. Then mes (Ω) shall be the (Lebesgue) measure of Re Ω only. The same shall hold for the complex domains named F and G (see next paragraph). The measure of the imaginary parts of complex domains is of no significance.

The following domains occur frequently in the proof: G is the complex domain of momenta p, and Ω is the complex domain of frequencies ω. The complex domain of coordinates q is $[0, 2\pi) \times [-i\rho, i\rho]$ since the q are 2π-periodic. F is the domain of the canonical variables (p, q), that is, $F = G \times ([0, 2\pi) \times [-i\rho, i\rho])$. Note that

$$\text{mes}(F) = (2\pi)^n \, \text{mes}(G). \tag{2.24}$$

3 Outline of the KAM Proof

This chapter discusses the main ideas and techniques used in Arnold's proof of the KAM theorem. Some of them are traditional, like perturbation theory, some were relatively new at Arnold's time, like super-convergence, sets without interior, or a Diophantine condition used to overcome problems with small divisors, and some were new, like the cutoff of the Fourier series.

There is a number of reviews on the KAM theorem and its proof. In order of increasing complexity and technical details, they may be roughly arranged as follows:

Dumas (2014) A beautiful book that contains historical and conceptual background material on the theorem. The ideas of the KAM proofs are described very clearly, with technical details largely left out.

Arnold et al. (1997) A section on the KAM theorem in a volume on modern theoretical mechanics. Also not many technical details.

Arnold (1989) A classical text on the general background of the theorem, like canonical transformations, symplectic manifolds, the Liouville theorem and Arnold's generalization of it. A general description of the KAM theorem, but not many technical details.

Thirring (1977) The first of the books and reviews that gives some details of the proof. Covers the KAM theorem (section 3.6) and treats perturbation theory, Fourier expansion, small divisors and a few norm estimates. 'A very pedagogical proof of a particular case of the result (that nevertheless contains the most essential difficulties)' (de la Llave 2001, p. 97).

Chierchia (2009) Encyclopedia article on the three proofs of the KAM theorem by Kolmogorov, Arnold and Moser. Discusses many aspects of more recent developments in KAM theory. Many technical details, notably in the Complementary Notes.

Arnold & Avez (1968) In appendix 34 Arnold gives details on the techniques applied in his proof, including some of the core lemmas. The treatment differs somewhat from that in Arnold (1963a,b).

de la Llave (2001) A tutorial on the KAM theory. Many of the more recent aspects, methods and techniques of the theory (cancellations in Lindstedt series; symplectic geometry; Whitney differentiability; implicit function theorems). However, regarding the classical KAM theory, 'since proofs of these theorems have been in the literature for several decades, and many of the estimates have been covered in the previous sections, we will leave many of the details to the reader, indicating the most interesting ones as exercises' (de la Llave 2001, p. 91).

Pöschel (2001) 'The paper is very well organized and written and can be used as a nice survey on KAM theory' (E. Valdinoci in Mathematical Reviews). A short yet detailed proof of the KAM theorem 'in its most basic form' (pp. 20-28).

DOI: 10.1201/9781003287803-3

Wayne (1996) Excellent review, similar in scope to Pöschel, perhaps with slightly more details on the convergence proof (Fundamental and Inductive lemma).

Rüssmann (1979) Focus is on theorems about uniform convergence of series with small divisors excluded by a Diophantine condition. The KAM theorem is proved in detail for the restricted three-body problem. The text is in German.

Sternberg (1969) Part II of his course on celestial mechanics deals with the KAM theorem and its proof, mostly in terms of circle maps. 'The main part of the book is Chapter 3, which is devoted to implicit function theorems of the type used by Nash for the imbedding of Riemannian manifolds in Euclidean space[1] and by Kolmogorov in celestial mechanics' (Arnold 1972).

The review by Broer & Sevryuk (2010) gives an excellent overview of recent developments in KAM theory and lists almost 500 papers and books on the subject. And there is a famous book by Siegel & Moser (1971), which treats Moser's proof of the KAM theorem in detail.

3.1 PERTURBATION SERIES VS. FAST ITERATION

In perturbation theory, unknown functions are expanded in power series with respect to a small parameter ε, and the equations of the problem are transformed into equations for the individual orders in ε. A well-known example of a perturbation series is that of particle (self-)interactions in quantum field theory using Feynman diagrams. In celestial mechanics, the gravitational force between planets leads to small corrections to the main attraction by the Sun. The Hamiltonian is then written with a perturbation term,

$$H(p,q) = H_0(p) + \varepsilon H_1(p,q), \qquad (3.1)$$

where H_0 is the Hamiltonian of the integrable Kepler problem and depends on the action variable $p = (p_1, \ldots, p_n)$ only, and εH_1 is the small perturbation including also the angle variable $q = (q_1, \ldots, q_n)$. H_1 is assumed to be 2π-periodic in each of the q_i. We consider a simple example of the perturbation series method introduced by Lindstedt, following the presentation in Arnold et al. (1997), p. 175. Let

$$S(P,q) = (P,q) + \varepsilon S_1(P,q) + \varepsilon^2 S_2(P,q) + \varepsilon^3 S_3(P,q) \qquad (3.2)$$

be the generating function of a canonical transformation to new canonical variables P, Q, where $(P,q) = \sum_i P_i q_i$. One wants to obtain a new Hamilton function K of the form, say,

$$K(P,Q) = K_0(P) + \varepsilon K_1(P) + \varepsilon^2 K_2(P) + \varepsilon^3 K_3(P) + \varepsilon^4 K_4(P,Q), \qquad (3.3)$$

where, since S does not depend on time explicitly,

$$H(p,q) = K(P,Q). \qquad (3.4)$$

---

[1] According to the Encyclopedia of Mathematics, a Euclidean space is a vector space $\mathbb{R}^n, n \in \mathbb{N}$ with inner product $(x,y) = \sum_{i=1}^n x_i y_i$ in Cartesian coordinates x_i, y_i for $x, y \in \mathbb{R}^n$. The points of an Euclidean space can be identified with vectors starting in the origin and ending in the point.

Abbreviating $S_{1,q} = \partial S_1/\partial q$, etc., the canonical transformation is given by (see Eq. 1.23)

$$p = S_q(P,q) = P + \varepsilon S_{1,q} + \varepsilon^2 S_{2,q} + \varepsilon^3 S_{3,q}, \tag{3.5}$$

$$Q = S_P(P,q) = q + \varepsilon S_{1,P} + \varepsilon^2 S_{2,P} + \varepsilon^3 S_{3,P}. \tag{3.6}$$

The second line will not be used in the following. $H(p,q)$ is written as

$$H(p,q) = H_0(P + \varepsilon S_{1,q} + \varepsilon^2 S_{2,q} + \varepsilon^3 S_{3,q}) + \varepsilon H_1(P + \varepsilon S_{1,q} + \varepsilon^2 S_{2,q}, q), \tag{3.7}$$

and a Taylor series expansion is performed about $H(P,q)$ up to terms in ε^3. For this reason, the term $\varepsilon^3 S_{3,q}$ in εH_1 was already left out. One obtains

$$
\begin{aligned}
H(p,q) = \; & H_0(P) \\
& + \varepsilon \left(H_{0,P}(P) S_{1,q}(P,q) + H_1(P,q) \right) \\
& + \varepsilon^2 \left(H_{0,P} S_{2,q} + H_{1,P} S_{1,q} + \tfrac{1}{2} H_{0,PP} S_{1,q} S_{1,q} \right) \\
& + \varepsilon^3 \left(H_{0,P} S_{3,q} + H_{1,P} S_{2,q} + H_{0,PP} S_{1,q} S_{2,q} + \right. \\
& \qquad \left. + \tfrac{1}{2} H_{1,PP} S_{1,q} S_{1,q} + \tfrac{1}{6} H_{0,PPP} S_{1,q} S_{1,q} S_{1,q} \right),
\end{aligned}
\tag{3.8}
$$

where we abbreviated

$$H_{0,PP} S_{1,q} S_{1,q} = \sum_{i=1}^{n} \sum_{j=1}^{n} H_{0,P_i P_j} S_{1,q_i} S_{1,q_j} \tag{3.9}$$

and so forth. Note especially that

$$H_{0,PP}(S_{1,q} S_{2,q} + S_{2,q} S_{1,q}) = 2 H_{0,PP} S_{1,q} S_{2,q} \tag{3.10}$$

by symmetry of $H_{0,P_i P_j}$. Identifying terms with the same power of ε in Eqs. (3.3) and (3.8) gives, in the form used by Arnold et al. (1997), p. 176,

$$
\begin{aligned}
K_0(P) &= H_0(P), \\
K_1(P) &= (\omega(P), S_{1,q}(P,q)) + H_1(P,q), \\
K_2(P) &= (\omega(P), S_{2,q}(P,q)) + F_2(P,q), \\
K_3(P) &= (\omega(P), S_{3,q}(P,q)) + F_3(P,q),
\end{aligned}
\tag{3.11}
$$

where the Hamilton equation $\dot{Q} = \partial H_0/\partial P = \omega(P)$ was used and $(\omega, S_{i,q})$ is the vector scalar product. It is not necessary here to further specify $F_2, F_3, \ldots$: Kolmogorov's suggestion was to *iterate* (repeat) the canonical transformation in the second lines of Eqs. (3.8) and (3.11), which include H_1 and $S_{1,q}$ only, instead of using a Lindstedt expansion to order > 1 in ε. Suppose that all functions are 2π-periodic in the q_i. (This is related to the averaging principle for rapid oscillations in celestial mechanics, for which we refer again to Arnold et al. 1997.) Fourier expansion[2] with respect

---

[2] '[S]ome recent proofs [of the KAM theorem] do not even use Fourier analysis.' (de la Llave 2001, p. 5)

to q gives

$$H_1(P,q) = \sum_k h_{1,k}(P) e^{i(k,q)},$$

$$F_j(P,q) = \sum_k f_{j,k}(P) e^{i(k,q)},$$

$$S_j(P,q) = \sum_k s_{j,k}(P) e^{i(k,q)}, \tag{3.12}$$

where the sums are over all tuples $k \in \mathbb{Z}^n$. This yields, for $j \geq 1$,

$$(\omega(P), S_{j,q}(P,q)) = \sum_{k \neq 0} i(k, \omega(P)) s_{j,k}(P) e^{i(k,q)}. \tag{3.13}$$

The missing Fourier component $k = 0$ is crucial here and in the following. In a sense, it causes half the effort in proving the KAM theorem, in that one has to handle the frequency map $A(p) = \omega$. Inserting Eq. (3.12) in Eq. (3.11) gives the equations

$$K_1(P) = \sum_{k \neq 0} i(k, \omega(P)) s_{1,k}(P) e^{i(k,q)} + h_{1,0}(P) + \sum_{k \neq 0} h_{1,k} e^{i(k,q)},$$

$$K_j(P) = \sum_{k \neq 0} i(k, \omega(P)) s_{j,k}(P) e^{i(k,q)} + f_{j,0}(P) + \sum_{k \neq 0} f_{j,k} e^{i(k,q)}. \tag{3.14}$$

The terms without a k-vector give

$$K_1(P) = h_{1,0}(P),$$

$$K_j(P) = f_{j,0}(P), \qquad j \geq 2. \tag{3.15}$$

Using orthogonality of the functions $e^{i(k,q)}$ in the k-sums gives

$$s_{1,k}(P) = \frac{i h_{1,k}(P)}{(k, \omega)},$$

$$s_{j,k}(P) = \frac{i f_{j,k}(P)}{(k, \omega)}, \qquad j \geq 2. \tag{3.16}$$

Inserting this in Eq. (3.12), one has

$$S_1(P,q) = s_{1,0}(P) + \sum_{k \neq 0} \frac{i h_{1,k}(P) e^{i(k,q)}}{(k, \omega(P))},$$

$$S_j(P,q) = s_{j,0}(P) + \sum_{k \neq 0} \frac{i f_{j,k}(P) e^{i(k,q)}}{(k, \omega(P))}, \qquad j \geq 2. \tag{3.17}$$

Equations (3.15) and (3.17) are the (formal) solution of the problem. The remarkable fact is that six functions $K_1, K_2, K_3, S_1, S_2, S_3$ were obtained from the last three equations in (3.11), due to the splitting of the Fourier components into $k = 0$ and $k \neq 0$. Equations (3.15) and (3.17) correspond to the set of equations after Eq. (26) in

Arnold et al. (1997), p. 176 (with the averaging and integration operators introduced on page 143). The functions $s_{1,0}$ and $s_{j,0}$ are arbitrary and, as Arnold et al. remark, 'one often takes' $s_{1,0} = s_{j,0} = 0$. The problem with Eq. (3.17) is the occurence of denominators $(k, \omega(P))$ that can become arbitrarily small at resonances. This is the central problem of celestial mechanics, and is overcome in the KAM theorem by excluding resonances via a Diophantine condition and using a fast iteration method, to which we turn now.

The equations from (3.11) onward remain valid if $j > 3$, i.e., for perturbation expansions up to any order ε^m,

$$K(P,Q) = K_0(P) + \varepsilon K_1(P) + \cdots + \varepsilon^m K_m(P) + \varepsilon^{m+1} K_{m+1}(P,Q). \tag{3.18}$$

The central aspect of Lindstedt's method is that, by going from perturbation order m to perturbation order $m + n$, the accuracy (with which the new Hamiltonian K resembles an integrable Hamiltonian) increases from ε^m to ε^{m+n}. Correspondingly, the time over which the trajectory is close to that of the integrable problem grows from $\approx 1/\varepsilon^m$ to $\approx 1/\varepsilon^{m+n}$. We consider now an *iteration* method instead of this series expansion. Performing then $m + n$ instead of m successive approximations, the accuracy increases from ε^{2^m} to $\varepsilon^{2^{m+n}}$. The basic idea is to consider not a single canonical transformation up to accuracy of ε^m as above, but instead a sequence of m canonical transformations, each with an accuracy increase of ε.

> As the works of A. N. Kolmogorov and V. I. Arnold revealed, this procedure of successive coordinate transformations actually possesses the remarkable property of quadratic convergence: following m transformations, the mismatch in the Hamiltonian depending on the phases has order ε^{2^m} (ignoring small denominators). This 'superconvergence' annihilates the effects of small denominators and ensures the convergence of the full procedure on a certain 'nonresonance' set. (Arnold et al. 1997, p. 180)

Or:

> The idea of Kolmogorov is to iterate this procedure [the Lindstedt series] also in the next approximations, so that instead of the usual sequence of approximations with accuracies $\varepsilon, \varepsilon^2, \varepsilon^3, \varepsilon^4, \ldots$ one obtains a fast converging sequence $\varepsilon, \varepsilon^2, \varepsilon^4, \varepsilon^8, \ldots$ (Arnold 1965, p. 100)[3]

The standard reference for an early method with super-convergence is Newton's tangent method (see below). One can include here other methods using successive approximations, e.g., the (Banach) contraction principle used in the proof of the implicit function theorem or in fixed point theorems (see Granas & Dugundji 2003) or

---

[3]I thank Prof. A. Pikovski for this translation from Russian.

Schmidt's method to solve non-linear integral equations (see Lichtenstein 1931). To start, assume again a slightly perturbed Hamiltonian,

$$H(p,q) = H_0(p) + \varepsilon H_1(p,q). \tag{3.19}$$

We aim at a canonical transformation to order ε only, thus Eqs. (3.2) and (3.3) become

$$S(P,q) = (P,q) + \varepsilon S_1(P,q),$$
$$K(P,Q) = K_0(P) + \varepsilon K_1(P) + \varepsilon^2 K_2(P,Q). \tag{3.20}$$

Inserting K_0 from Eq. (3.11) and K_1 from Eq. (3.15) in $H(p,q) = K(P,Q)$ gives

$$H_0(p) + \varepsilon H_1(p,q) = H_0(P) + \varepsilon h_{1,0}(P) + \varepsilon^2 K_2(P,Q), \tag{3.21}$$

and $S_1(P,q)$ is given again by Eq. (3.17).

> The new Hamiltonian has the same form as the old one, but the phases $[Q]$ appear only in the terms of order ε^2. We now subject the system we obtained to a similar coordinate transformation. This 'pushes' the phases further to the terms of order ε^4. After m such coordinate transformations the dependence of phases is preserved only in the terms of order ε^{2^m}. We remind the reader that in Lindstedt's method, after the m-th approximation coordinate transformation, the dependence on the phases is 'pushed' to the terms of order ε^{m+1} only. (Arnold et al. 1997, p. 181)

With $\varepsilon' = \varepsilon^2$, write Eq. (3.21) for $H_0(p) + \varepsilon H_1(p,q)$ as

$$K(P,Q) = [H_0(P) + \varepsilon h_{1,0}(P)] + \varepsilon' K_2(P,Q). \tag{3.22}$$

As stated in the quotation above, this is of the same form as Eq. (3.19), thus a further canonical transformation can be applied to obtain a Hamiltonian

$$L(P,Q) = [H_0(P) + \varepsilon h_{1,0}(P)] + \varepsilon' k_{2,0}(P) + \varepsilon'^2 L_2(P,Q), \tag{3.23}$$

where $k_{2,0}(P)$ is the Fourier component $k = 0$ of $K_2(P,Q)$, as $h_{1,0}(P)$ is the component $k = 0$ of $H_1(P,q)$. Thus

$$L(P,Q) = [H_0(P) + \varepsilon h_{1,0}(P) + \varepsilon^2 k_{2,0}(P)] + \varepsilon^4 L_2(P,Q), \tag{3.24}$$

and so forth, with the approximation error of the last term containing the phases falling off as $\varepsilon \to \varepsilon^2 \to \varepsilon^4 \to \varepsilon^8 \ldots$ instead of $\varepsilon \to \varepsilon^2 \to \varepsilon^3 \to \varepsilon^4 \ldots$ for the original Lindstedt series.[4] This is called quadratic convergence or super-convergence. Still, since the two methods do not appear so different at first sight, the following is not surprising:

---

[4] 'In the study of Lindstedt series [...], the proof of convergence by exhibiting explicitly cancellations of the series was accomplished in Eliasson (1996). Contrary to the terms in the expansions considered in Siegel (1942), the terms in the Lindstedt series do grow very fast and one cannot establish convergence by just bounding sizes but one needs to exhibit cancellations in the terms.' (de la Llave 2001, p. 6)

The procedure of successive coordinate changes was proposed by Newcomb. We owe its present form to Poincaré, who, however, considered Newcomb's procedure to be equivalent to that of Lindstedt. (Arnold et al. 1997, p. 180)

Kolmogorov's idea was to use a method with quadratic convergence to overcome problems with small divisors. The following must however be noted:

The estimate ε^{2^m} indicates the formal order in ε of the mismatch in the Hamiltonian. Actually the mismatch can be considerably larger due to the influence of small denominators. (Arnold et al. 1997, p. 181)

The traditional 'method with quadratic convergence' is Newton's tangent method, used to find the zeros of a real-valued function $f(x)$. This is treated at some length in Rüssmann (1979) and Sternberg (1969) and in texts on numerical mathematics. We do not go into this here, but Fig. 3.1 merely shows the converge of three algorithms solving the equation $(x-1)^3 = 0$ for its root $x = 1$, starting at $x = 1.5$. The three methods used are (i) the Newton scheme,

$$x_{n+1} = x_n - \frac{f(x_n)}{f'(x_n)}, \tag{3.25}$$

(ii) the regula falsi,

$$x_{n+1} = x_n - f(x_n)\,\frac{x_n - x_{n-1}}{f(x_n) - f(x_{n-1})} \tag{3.26}$$

and (iii) a switch method without slopes that requires at start an interval $x_- < x_0 < x_+$ with $f(x_+)/f(x_-) < 0$,

$$x_{n+1} = (x_- + x_+)/2,$$
$$\text{if } f(x_{n+1})/f(x_-) > 0 \text{ then } x_- = x_{n+1} \text{ else } x_+ = x_{n+1}. \tag{3.27}$$

The latter simple and robust (no slopes) method beats the Newton method clearly. For a derivation of the convergence rates of the Newton method and the regula falsi, 2 and $(1+ \sqrt{5})/2$, respectively, see Rüssmann (1979).

The case at hand, i.e., the canonical transformation of the perturbed Hamiltonian to new variables so that the perturbation becomes 'quadratically small' is a problem from functional analysis, not of real analysis (functions are iterated, not variables). The following lemma for such a setting and its proof are taken from Arnold & Avez (1968), p. 250. Almost any of the following lines reappears in the Fundamental theorem of the next chapter. The essential equation is (3.28), which corresponds to the quadratic convergence of Newton's method. The denominator δ^v are terms that counteract this convergence, and will also appear in the following.

Iteration lemma *Given are*
 – arbitrary real constants $v > 0$ and $d > 0$,

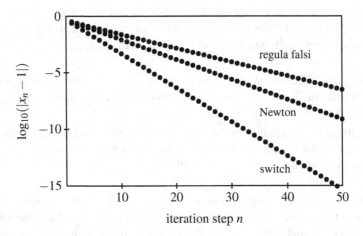

Figure 3.1 Convergence of the Newton algorithm, the regula falsi and a switch algorithm to the zero of $f(x) = (x-1)^3$

– a complex function $f(z)$ analytic on a domain $G \subset \mathbb{C}$,

– an operator L that maps f to a function $L[f]$.[5]

Assume that for all $0 < \delta < d$ the function $L[f]$ is analytic in $G - \delta$ and

$$\|L[f]\|_{G-\delta} < \frac{\|f\|_G^2}{\delta^v}. \tag{3.28}$$

Then for all $\delta > 0$ there exists an $\varepsilon(\delta) > 0$ such that

$$\|f\|_G < \varepsilon \qquad \text{implies} \qquad \sum_{s=0}^{\infty} \|L^s[f]\|_{G-\delta} < 2\varepsilon. \tag{3.29}$$

Remark Here $L^0[f] = f$, $L^2[f] = L[L[f]]$, etc. The variable ε will be renamed to M in the KAM proof.

Proof For given $\delta > 0$ choose

$$\delta_1 < \min\left\{\frac{1}{8}, \frac{\delta}{2}\right\}. \tag{3.30}$$

Define two infinite sequences,

$$\delta_{s+1} = \delta_s^{3/2},$$
$$\varepsilon_s = \delta_s^{2v+1}. \tag{3.31}$$

---

[5] Arnold & Avez (1968) give as example $L[f] = f f'$ with complex derivative f'.

Then also
$$\varepsilon_{s+1} = \delta_{s+1}^{2\nu+1} = (\delta_s^{3/2})^{2\nu+1} = (\delta_s^{2\nu+1})^{3/2} = \varepsilon_s^{3/2}. \tag{3.32}$$

One easily shows that (see Eqs. 4.136 and 4.137)

$$\sum_{s=1}^{\infty} \delta_s < 2\delta_1 < \delta, \tag{3.33}$$

$$\sum_{s=1}^{\infty} \varepsilon_s < 2\varepsilon_1. \tag{3.34}$$

Define an infinite sequence of domains, $s = 1, 2, \ldots,$

$$G_1 = G, \qquad\qquad G_{s+1} = G_s - \delta_s. \tag{3.35}$$

Then for $G - \delta$ in the statement of the lemma, because $\delta > \delta_1 + \cdots + \delta_s$,

$$G - \delta \subset G - \delta_1 - \cdots - \delta_s = G_{s+1} \tag{3.36}$$

for all s. Assume that the antecedent in Eq. (3.29) is true with $\varepsilon = \varepsilon_1(\delta_1)$ (and we have put $G = G_1$),

$$\|f\|_{G_1} < \varepsilon_1. \tag{3.37}$$

Using this as induction start, we show that for all $s = 1, 2, \ldots,$

$$\left\| L^{s-1}[f] \right\|_{G_s} < \varepsilon_s. \tag{3.38}$$

Assume thus $\left\| L^{s-1}[f] \right\|_{G_s} < \varepsilon_s$. Then (replacing f in 3.28 by $g = L^{s-1}[f]$, which is allowed since $L^s[f]$ is analytic in $G_s - \delta_s$)

$$\|L^s[f]\|_{G_{s+1}} = \left\| L[L^{s-1}[f]] \right\|_{G_s - \delta_s} \overset{(3.28)}{<} \frac{\left\| L^{s-1}[f] \right\|_{G_s}^2}{\delta_s^\nu} <$$

$$< \frac{\varepsilon_s^2}{\delta_s^\nu} = \delta_s^{3\nu+2} = \delta_s^{1/2}(\delta_s^{3/2})^{2\nu+1} < \delta_{s+1}^{2\nu+1} = \varepsilon_{s+1}, \tag{3.39}$$

which finishes the induction step. Using now Eq. (3.38), one has

$$\sum_{s=0}^{\infty} \|L^s[f]\|_{G-\delta} \le \sum_{s=1}^{\infty} \left\| L^{s-1}[f] \right\|_{G_s} < \sum_{s=1}^{\infty} \varepsilon_s < 2\varepsilon_1 = 2\varepsilon, \tag{3.40}$$

where the '$\le$' holds since the domain $G - \delta$ is smaller than each G_s by Eq. (3.36), thus gives the smallest norm for a given f. $\qquad\qquad\qquad\qquad\qquad\qquad\qquad \square$

3.2 PERTURBATION THEORY WITH NEW FREQUENCIES

Arnold (1963a,b) rewrites the infinitesimal canonical transformation of the last section in a form that is used in his Fundamental theorem of the next chapter. An emphasis is on the frequency shift caused by the integrable part of the perturbed Hamiltonian. We start with a single degree of freedom, $n = 1$, and generalize later to arbitrary

n. Instead of H_0 and H_1 from the last section we write now and in all the following H^+ and H^-, reserving a subscript s to enumerate canonical transformations. Let $H^+(p)$ be the unperturbed Hamiltonian of an integrable problem with action-angle variables p and q. The angular frequencies of motion on a phase-space torus are given by

$$\omega = A(p) = \frac{\partial H^+(p)}{\partial p}(=\dot{q}). \tag{3.41}$$

The Hamiltonian shall be subject to a small perturbation,

$$H(p,q) = H^+(p) + \varepsilon H^-(p,q) \tag{3.42}$$

with $\varepsilon \ll 1$. An infinitesimal canonical transformation $(p,q) \mapsto (P,Q)$ shifts this to order ε^2,

$$p = P + \varepsilon \frac{\partial}{\partial q} S(P,q), \tag{3.43}$$

$$Q = q + \varepsilon \frac{\partial}{\partial P} S(P,q). \tag{3.44}$$

Arnold (1963a,b) introduces the identity

$$\begin{aligned}
H(p,q) = &\, H^+(P) + \varepsilon H^-(P,q) \\
&+ H^+(p) - H^+(P) \\
&+ \varepsilon\big(H^-(p,q) - H^-(P,q)\big).
\end{aligned} \tag{3.45}$$

Note that $H^-(P,q)$ is used here, not $H^-(P,Q)$. With $p-P$ from Eq. (3.43), this can be written

$$\begin{aligned}
H(p,q) = &\, H^+(P) + \varepsilon H^-(P,q) \\
&+ \frac{H^+(p) - H^+(P)}{p-P}\, \varepsilon \frac{\partial}{\partial q} S(P,q) \\
&+ \varepsilon \frac{H^-(p,q) - H^-(P,q)}{p-P}\, \varepsilon \frac{\partial}{\partial q} S(P,q).
\end{aligned} \tag{3.46}$$

The last line is $\sim \varepsilon^2$, and treated as 'term of higher order.' In the next chapter, however, these terms have to be evaluated carefully to obtain upper bounds for a convergence proof. With Eq. (3.41), the fraction in the middle line can be approximated by $\omega = \partial H^+(p)/\partial p$ or $\omega' = \partial H^+(P)/\partial P$, the difference $\omega - \omega'$ being of order ε, thus leading to another term of order ε^2. Therefore,

$$H(p,q) = H^+(P) + \varepsilon \left[H^-(P,q) + \omega \frac{\partial}{\partial q} S(P,q) \right] + \varepsilon^2 \dots, \tag{3.47}$$

cf. Eq. (3.11). The analytic functions $H^-(P,q)$ and $S(P,q)$ are 2π-periodic in q and can be expanded in Fourier series,

$$H^-(P,q) = h_0(P) + \sum_{k\in\mathbb{Z}\backslash\{0\}} h_k(P)\,e^{ikq},$$

$$S(P,q) = s_0(P) + \sum_{k\in\mathbb{Z}\backslash\{0\}} s_k(P)\,e^{ikq}, \qquad (3.48)$$

where $s_0(P)$ drops out from Eq. (3.46). The term $h_0(P)$, however, plays a major role in the KAM theorem; it gives the momentum shift and a corresponding frequency shift caused by the perturbation. When resonant frequencies ω are excluded by a Diophantine condition (see below), the perturbed frequencies ω' may be resonant again. Lemma G5 in Section 6.9 defines a diffeomorphism $P \mapsto \omega'$ on a restricted P-domain so that the ω' are non-resonant. Inserting Eq. (3.48) in Eq. (3.47) gives

$$H(p,q) = H^+(P) + \varepsilon h_0(P) + \varepsilon \sum_{k\in\mathbb{Z}\backslash\{0\}} \left[h_k(P) + ik\omega s_k(P)\right]e^{ikq} + \varepsilon^2\ldots, \qquad (3.49)$$

where from the Fourier theorem,

$$h_0(P) = \frac{1}{2\pi}\int_0^{2\pi} dq\, H^-(P,q) =: \overline{H}(P). \qquad (3.50)$$

For the shift from $H^+(p)$ to $H^+(P) + \varepsilon h_0(P)$, see also Eqs. (3.22) and (3.24). We demand that the k-sum in Eq. (3.49) vanishes. Since the e^{ikq} form an orthogonal system, each of the square brackets in Eq. (3.49) has to vanish individually, giving again

$$s_k(P) = \frac{ih_k(P)}{k\omega}. \qquad (3.51)$$

The case $n > 1$ is treated correspondingly and one obtains,[6] for vectors $q = (q_1,\ldots,q_n)$, $P = (P_1,\ldots,P_n)$ and $k = (k_1,\ldots,k_n)$,

$$\overline{H}(P) = \frac{1}{(2\pi)^n}\int_0^{2\pi} dq_1 \ldots \int_0^{2\pi} dq_n\, H^-(P,q) \qquad (3.52)$$

and

$$s_k(P) = \frac{ih_k(P)}{(k,\omega)}. \qquad (3.53)$$

From now on, the zeroth Fourier component $\overline{H}(p)$ in Eqs. (3.50) resp. (3.52) is split off from the perturbed Hamiltonian,

$$\tilde{H}(p,q) = H^-(p,q) - \overline{H}(p), \qquad (3.54)$$

where

$$\int_0^{2\pi} dq_1 \ldots \int_0^{2\pi} dq_n\, \tilde{H}(p,q) = 0. \qquad (3.55)$$

---

[6]The following formula is given in Arnold (1963b), p. 97, but not in Arnold (1963a).

We follow Arnold (1963a) and write this latter equation as

$$\oint dq\, \tilde{H}(p,q) = 0. \tag{3.56}$$

Despite its appearance, this has nothing to do with Cauchy's integral theorem. The frequency shift by the perturbation is $\omega' - \omega = \partial \overline{H}(p)/\partial p$. Note that $\partial \tilde{H}(p,q)/\partial p = 0$. Furthermore, $\|H^-\| \leq M$ implies $\|\overline{H}\| \leq M$ by Eqs. (3.50) or (3.52). Equation (3.54) and the triangle inequality imply then $\|\tilde{H}\| \leq 2M$.

Both the canonical transformation of $H(p,q)$ to a new Hamiltonian with smaller norm and the corresponding frequency shift due to $\overline{H}(p)$ are treated in the Fundamental theorem of the next chapter. The infinite sequence of canonical transformations that brings the perturbed Hamiltonian to integrable form, $H_\infty(p_\infty)$, is subject of the Inductive theorem. The iteration number is indicated by a subscript $s = 1,2,\ldots$ to the relevant quantities. The original canonical variables and perturbed Hamiltonian at start receive a subscript 0, thus $H(p,q) = H_0(p_0,q_0) = H_0^+(p_0) + H_0^-(p_0,q_0)$. The sequence of transformations of the Hamiltonian is then as follows:[7]

$$
\begin{aligned}
H(p,q) &= H_0^+(p_0) + H_0^-(p_0,q_0) \\
&= H_0^+(p_0) + \overline{H}_0(p_0) + \tilde{H}_0(p_0,q_0) \\
&\overset{}{=} \quad\quad H_1^+(p_0) + \tilde{H}_0(p_0,q_0) \\
&\overset{*}{=} \quad\quad H_1^+(p_1) + H_1^-(p_1,q_1) \\
&= \quad\quad H_1^+(p_1) + \overline{H}_1(p_1) + \tilde{H}_1(p_1,q_1) \\
&= \quad\quad\quad\quad H_2^+(p_1) + \tilde{H}_1(p_1,q_1) \\
&\overset{*}{=} \quad\quad\quad\quad H_2^+(p_2) + H_2^-(p_2,q_2),
\end{aligned}
\tag{3.57}
$$

and so forth. Since the canonical transformations are independent of time, $H_{s-1}^+(p_{s-1}) + H_{s-1}^-(p_{s-1},q_{s-1}) = H_s^+(p_s) + H_s^-(p_s,q_s)$ for all s. A reduction from three to two sum terms in Eq. (3.57) indicates the frequency shift in H^+ due to $\overline{H}$, and the equations marked with '$*$' refer to the canonical transformation. Note that the appearance of old angles in the new Hamiltonian, $H^-(P,q)$, is an artefact of Eqs. (3.11) and (3.47). These terms vanish exactly by the very assumption on the canonical transformation. The new perturbed Hamiltonian H_s^- (of order $\varepsilon^2, \varepsilon^4, \varepsilon^8, \ldots$) in Eq. (3.57) contains the new variables p_s, q_s.

3.3 FREQUENCY DIFFEOMORPHISM

[...] during evolution the frequencies $\omega(I)$ themselves vary slowly. As a result, over time $1/\varepsilon$, the phase point may repeatedly visit the neighborhood

---

[7]Arnold (1963a,b) uses a simpler notation, which, however, requires some renaming of quantities in the two main theorems.

of the resonant surfaces. Consequently, even the first-approximation coordinate change [infinitesimal canonical transformation] is not defined, in general, along the entire trajectory over time $1/\varepsilon$. The indicated coordinate change nevertheless remains the main tool for analysing the motion between resonances. (Arnold et al. 1997, p. 144)

To overcome this obstacle, Kolmogorov suggested the following approach.

Next, near a non-resonant torus of the unperturbed system corresponding to a fixed value of the frequencies, we will look for an invariant torus of the perturbed system on which there is conditionally-periodic motion with exactly the same frequencies as the ones we fixed, and which necessarily satisfy the condition of being non-resonant described above. In this way, instead of the variations of frequency customary in perturbation schemes (consisting of the introduction of frequencies depending on the perturbation), we must hold constant the non-resonant frequencies, while selecting initial conditions depending on the perturbation in order to guarantee motion with the given frequencies. This can be done by a small (when the perturbation is small) change of initial conditions, because the frequencies change with the action variables according to the non-degeneracy condition. (Arnold 1989, p. 405)

This last quotation addresses the fact that the constant frequencies ω^* refer to different initial conditions $p^* = A^{-1}(\omega^*)$ and $p'^* = A'^{-1}(\omega^*)$ for the unperturbed and perturbed system, respectively. Thus, the equation (see Eq. 3.47)

$$H^-(P,q) + (\omega(P), S_q(P,q)) = 0 \qquad (3.58)$$

is replaced by

$$H^-(P,q) + (\omega^*, S_q(P,q)) = 0 \qquad (3.59)$$

with constant ω^*. The difference between these two equations is, using $S_q = p - P$,

$$(\omega(P) - \omega^*(p^*), p - P). \qquad (3.60)$$

For $\omega(P)$ not too steep (controlled by the bound $\Theta > 1$ in the following), this scales with the square in momentum differences, and is small of second order, see Arnold (1963b), p. 99. Arnold takes a radically different approach here:

Returning to the classical ideas of the theory of perturbations *we disregard* [the idea to look for only one invariant torus T_{ω^*}] *and take* ω [...] *to be a function* $\omega(p) = \partial H_0/\partial p$. (Arnold 1963b, pp. 104-105)

And:

Rather than trying to perform a change of variables that produces one torus, the method of Arnold (1963a) produces changes of variables that reduce the system to approximately integrable in a region of space. Hence, the method of Arnold (1963a) produces all the tori at the same time. (de la Llave 2001, p. 99)

3.4 DIOPHANTINE CONDITION

A Diophantine condition is used to exclude frequencies (of motion on a phase-space torus) that are in resonance, $(k, \omega) \approx 0$, and thus cause small divisors.

It is not difficult to see that, unless we impose some quantitative restriction on how fast $|k \cdot \omega|^{-1}$ can grow, the solutions given by [3.53] may fail to be even distributions. (de la Llave 2001, p. 26; our equation number)

With small divisors thus excluded, Fourier series converge uniformly on the remaining set of large measure of non-resonant frequencies. The method was first applied by Siegel (1942), then suggested (apparently independent) by Kolmogorov (1954) to overcome small-divisor problems in classical mechanics, and put to full use by Arnold (1961, 1963a,b) and Moser (1962). An early book on Diophantine approximations – a branch of number theory – is Minkowski (1907), a more recent collection of papers on the subject is Hlawka (1990), which contains also a scientific biography of C. L. Siegel. There is a connection between Diophantine approximations and Fermat's 'Great theorem' (Mordell's conjecture, proved by Faltings 1983), see the 'editorial comments' after Voronin (1995), p. 287, or the lecture by Vojta (1993).

For two vectors $\omega \in \mathbb{R}^n$ and $k \in \mathbb{Z}^n$, the Diophantine condition used in the KAM theorem is

$$|(k, \omega)| \geq \frac{K}{|k|^\nu}, \tag{3.61}$$

where $|k| = |k_1| + \cdots + |k_n|$ and $\nu = n + 1$. $K > 0$ is a free constant.

Lemma *For $K \to 0$ the total measure of resonance zones in which the Diophantine condition (8.27) is violated tends to zero.*

For the *proof*, see Arnold (1963b), p. 98; Arnold & Avez (1968), p. 252; Pöschel (2001), p. 5; Rüssmann (1979), p. 197; de la Llave (2001), p. 33. The following is a related theorem for $n = 1$:

[F]or almost-all α there exists an infinite number of rational approximations a/q satisfying the inequality

$$\left| \alpha - \frac{a}{q} \right| < \frac{1}{q^2 \ln q}, \tag{3.62}$$

whereas the inequality

$$\left| \alpha - \frac{a}{q} \right| < \frac{1}{q^2 (\ln q)^{1+\varepsilon}}, \tag{3.63}$$

has for any $\varepsilon > 0$ an infinite number of solutions only for a set of numbers α of measure zero. (Sprindzhuk 1995, p. 279)

For resonances with small $|k|$, rather broad frequency bands or strips are excluded, while for resonances with large $|k|$, the excluded bands are narrow. Thus, the excluded strips at large $|k|$ may fully or partially fall into those excluded at small $|k|$.

The following material is taken from Lemma 34.7 in Arnold & Avez (1968), p. 252. For similar considerations, see Rüssmann (1979), pp. 185-197, and Wayne (1996), pp. 7-8. Consider the model equation (cf. Eq. 3.27 in de la Llave 2001, p. 36)

$$S(q+\omega) - S(q) = H(q), \qquad (3.64)$$

where H is given and S is unknown. Momentum p is suppressed since it appears as a constant parameter only. H and S shall be 2π-periodic in the angle variables q_1 to q_n. A Fourier decomposition gives

$$H(q) = \sum_{k\neq0} h_k e^{i(k,q)},$$

$$S(q) = \sum_{k\neq0} s_k e^{i(k,q)},$$

$$S(q+\omega) = \sum_{k\neq0} s_k e^{i(k,q)} e^{i(k,\omega)}. \qquad (3.65)$$

The term $k = 0$ is excluded from the present considerations (see above). Inserting Eq. (3.65) in Eq. (3.64) gives

$$\left(e^{i(k,\omega)} - 1\right) s_k = h_k, \qquad (3.66)$$

thus

$$s_k = \frac{h_k}{e^{i(k,\omega)} - 1} \qquad (3.67)$$

and

$$S(q) = \sum_{k\neq0} \frac{h_k}{e^{i(k,\omega)} - 1} e^{i(k,q)}. \qquad (3.68)$$

This is only a formal solution, since the convergence of the sum and thus the existence of S remains doubtful for $(k,\omega) \to 0$. The following lemma states convergence of this sum if a Diophantine condition is used.

Lemma *Let $H(q)$ be analytic and 2π-periodic and $|H(q)| \leq M$ for $|\mathrm{Im}\, q| \leq \rho$. Let*

$$\left|e^{i(k,\omega)} - 1\right| \geq \frac{K}{|k|^{n+1}} \qquad (3.69)$$

for given $K > 0$ (Diophantine condition). Then $S(q)$ from Eq. (3.64) as given by Eq. (3.68) is analytic and

$$|S(q)| < \frac{MC(n)}{K\delta^{2n+1}} \qquad \text{for} \qquad |\mathrm{Im}\, q| \leq \rho - 2\delta, \qquad (3.70)$$

where $C(n)$ depends only on n and δ is a small number with $\rho - 2\delta > 0$.

Remark The denominator δ^{2n+1} in Eq. (3.70) and the domain restriction to $|\mathrm{Im}\, q| \leq \rho - 2\delta$ are directly related. Wayne (1996), p. 8, writes formulas of this type suggestively as $\|S\|_{\rho-\delta} < C/\delta^{\nu}$. The statement of the lemma is discussed in virtually any presentation of KAM theory, see Dumas (2014), pp. 63-66.

Proof We use Lemma A2 on Fourier coefficients from Section 5.3:

(I) *If* $|H(q)| \leq M$ *for* $|\mathrm{Im}\, q| \leq \rho$ *then* $|h_k| \leq M e^{-|k|\rho}$.

(II) *If* $|s_k| \leq M' e^{-|k|\sigma}$ *then* $|S(q)| < \dfrac{M' C_1(n)}{\delta^n}$ *for* $|\mathrm{Im}\, q| \leq \sigma - \delta$,

where C_1 is a function of n only. We need a formula to replace powers by exponentials. For sufficiently large x, e^x is larger than any x^n for given n; so for any x^n one has $x^n < C_2(n-1) e^x$ for $x > 0$ and some function $C_2(n-1)$ (which is determined in Lemma A1). Putting $x = |k|\delta$, this gives

$$|k|^{n+1} < C_2(n) \frac{e^{|k|\delta}}{\delta^{n+1}}. \tag{3.71}$$

Then for the Fourier coefficients s_k from Eq. (3.67),

$$|s_k| = \frac{|h_k|}{|e^{i(k,\omega)} - 1|} \overset{(\mathrm{I})}{\leq} \frac{M e^{-|k|\rho}}{|e^{i(k,\omega)} - 1|} \overset{(3.69)}{\leq} M e^{-|k|\rho} \frac{|k|^{n+1}}{K}$$

$$\overset{(3.71)}{<} \frac{M}{K} e^{-|k|\rho} C_2(n) \frac{e^{|k|\delta}}{\delta^{n+1}} = \frac{M C_2(n)}{K \delta^{n+1}} e^{-|k|(\rho-\delta)}. \tag{3.72}$$

Applying (II) with $M' = M C_2(n)/(K\delta^{n+1})$ and $\sigma = \rho - \delta$ gives, for $|\mathrm{Im}\, q| \leq \sigma - \delta = \rho - 2\delta$,

$$|S(q)| < \frac{M C_1(n) C_2(n)}{K \delta^{2n+1}}, \tag{3.73}$$

which is Eq. (3.70).[8] Since the Fourier coefficients s_k in Eq. (3.72) fall off exponentially with $|k|$, $S(q)$ is analytic (see Lemma A2).   □

3.5   FAST ITERATION PLUS DIOPHANTINE CONDITION

With the method of the last section that led to Eq. (3.73), i.e., avoiding small (resonance) divisors by using a Diophantine condition, we can obtain an analytic generating function $S_s : (P,q) \mapsto (p,Q)$ of a single infinitesimal transformation and from this the corresponding canonical diffeomorphism $B_s : (P,Q) \mapsto (p,q)$.

---

[8] For one degree of freedom (circle diffeomorphisms) with $k = (k_1)$, one can use the complete Gamma function, $\Gamma(n+2) = \int_0^\infty dk\, e^{-k} k^{n+1}$, to majorize the sum $\sum_{k \neq 0} e^{-k(\rho-\delta)} k^{n+1}$ in the calculation leading to Eq. (3.72), see Wayne (1996), p. 8, instead of using Eq. (3.71). With a cutoff N of the Fourier series this also works for $n > 1$, since one has then an upper limit for the number of k-vectors with given $|k| = N$.

However, (1) the functions so obtained depend on p everywhere discontinu-
ously; (2) the convergence of all the approximations p_s, q_s as $s \to \infty$ remains
doubtful. (Arnold 1963b, p. 99)

Item (1) is addressed in the next section on resonance cutoff, item (2) is considered here. We have so far set-up a canonical transformation to new variables and used
a Diophantine condition to obtain analytic S. It remains doubtful, however, whether
repeated applications of these transformations leave finite domains, despite contin-
uing domain reductions. Indeed, the super-convergent iteration scheme guarantees
convergence of $S_1 \circ S_2 \circ S_3 \circ \cdots$ on finite domains:

> The principal point which we must check is that we don't lose all of our do-
> main of analyticity as we go through the argument note that $\tilde{\phi}$ [corresponding
> here to the generating function S] is analytic on a narrower strip than was our
> original diffeomorphism ϕ. The essential reason that there is a nonvanishing
> domain of analyticity at the completion of the argument is that the amount by
> which the analyticity strip shrinks at the nth step in the induction will be pro-
> portional to the amount by which our diffeomorphism differs from a rotation
> [here: the identity map] at the nth iterative step, and thanks to the extremely
> fast convergence of Newton's method, this is very small. (Wayne 1996, p. 11)

Newton's (or Newcomb's) method has thus two central aspects. First, for arbitrary
$\rho_0 > 0$ (see Eqs. 4.110, 4.111, 4.122 in the Inductive theorem), one has

$$|\text{Im } q_\infty| < \rho_\infty \qquad \text{with} \qquad \rho_\infty = \rho_0 - 3\gamma_1 - 3\gamma_2 - 3\gamma_3 - \cdots > \frac{\rho_0}{3}, \qquad (3.74)$$

and similarly for other domains; and second,

$$\left\| H_s^- \right\| \leq M_{s+1} = \frac{M_s^2}{\delta_s^\nu}. \qquad (3.75)$$

We shall now show, with [Eq. (3.75)] at our disposal, how to construct con-
verging successive approximations to the invariant torus. (Arnold 1963b, p.
99)

The main goal of this application is to show in action perhaps the most basic
heuristic principle of the KAM method: *Quadratic convergence can overcome
small divisors.* (de la Llave 2001, p. 60)

Or,

> [...] what we mean by smaller error is that the size of the new error will be
> bounded (in a smaller domain than the original one) by the square of the size of
> the original error times a factor that is the domain loss parameter to a negative
> power. (de la Llave 2001, p. 110)

The very small number δ_s^ν in the denominator of Eq. (3.75), counteracting the gain from $M_s \to M_{s+1} \sim M_s^2$, is due to (i) small divisors at a resonance (giving δ_s^{n+1}) and (ii) the requirement that S_s is analytic (giving another δ_s^n), see the lemma above. That the powers of δ_s in the denominator do not destroy the convergence of the fast iteration method was shown in Eq. (3.29) of the Iteration lemma in Section 3.1. In the Iteration lemma, convergence was achieved by making $\varepsilon_s = \delta_s^{2\nu+1}$ in Eq. (3.31) a sufficiently large power of δ_s, with δ_s^ν appearing in the denominator of the quadratic-convergence condition (3.28). Similarly, convergence of Eq. (3.75) is achieved by putting $M_s = \delta_s^T$ with sufficiently large T. Arnold (1963a) chooses $T = 8n+24$. The denominator in Eq. (3.75) will be found to be $\beta_s^2 \delta_s^{2\nu}$ with $\beta_s = \delta_s^3$ and $\nu = 2n+3$; thus δ_s^{4n+12} appears in the denominator of Eq. (3.75). With this,

$$\left\| H_s^- \right\| \leq M_{s+1} = \frac{\delta_s^{16n+48}}{\delta_s^{4n+12}} = \delta_s^{12n+36} = \left(\delta_s^{8n+24}\right)^{3/2} = M_s^{3/2}. \tag{3.76}$$

Thus for $M_1 < 1$ one has $M_s \to 0$ for $s \to \infty$. (For a similar consideration compare especially Eqs. 4.32 and 4.33 in de la Llave 2001, p. 67.) We have therefore found that a Diophantine condition together with a fast iteration scheme solve the problem of small divisors:

> Inverse powers of δ_j [...] appear in the smallness conditions [...]: this 'divergence' will, however, be beaten by the superfast decay of ε^{2^j}. (Chierchia 2009, p. 5073)

> By the rapid convergence of the Newton scheme [...] the error term converges to zero even faster [than the unbounded terms], thus allowing to overcome this effect of the small divisors. (Pöschel 2001, p. 19)

However:

> The fact that the convergence does indeed take place [for a scheme with quadratic convergence] is rather subtle. In our opinion, the only way to appreciate the subtlety of the convergence achieved by KAM theory is to give a serious try to several other seemingly reasonable schemes and see them fail. (de la Llave 2001, p. 60)

3.6 RESONANCE CUTOFF

> [I]n any neighbourhood of any invariant torus of the unperturbed system there is an n-dimensional torus on which all the trajectories after a certain time become closed. (Arnold 1963b, p. 97)

This means, the resonances (closed trajectories) lie dense.

[T]he frequency $H_p(P)$ is *a function of the action P* and since [...] $p \to H_p(p)$ is a local diffeomorphism, it follows that, in any neighborhood of p_0, there are points p such that $(k, H_p(p)) = 0$ for some $k \in \mathbb{Z}^n$. Thus, in any neighborhood of p_0, some divisors [...] will actually vanish and, therefore, *an analytic solution S cannot exist.* (Chierchia 2009, p. 5076)

Arnold's key idea to solve this problem is to consider, for each canonical transformation B_s, only resonances with $|k| < N_s$ where $N_s \gg 1$. Then for any non-resonant frequency point there is a whole (small but finite) neighborhood of non-resonant frequency points for which the Diophantine condition holds. The following three quotations catch the idea.

[...] in order not to have to deal with everywhere discontinuous functions of *p, we restrict ourselves in the series of the theory of perturbations to a finite number of harmonics* [...] (Arnold 1963b, p. 105)

It is absolutely essential [...] that ω satisfies infinitely many small divisor conditions, thus restricting ω to a Cantor set with no interior points. On the other hand, we will also need to transform the frequencies and thus want them to live in open domains. This conflict is resolved by approximating P [an infinite Fourier series] by a trigonometric polynomial R. Then only finitely many Fourier coefficients need to be considered at each step, and only finitely many small divisor conditions need to be required, which are easily satisfied on some open ω-domain. Of course, during the iteration more and more conditions have to be satisfied, and in the end these domains will shrink to some Cantor set. (Pöschel 2001, p. 19)

Having S be defined only on the complement of a dense set of points $\hat{p}$ [momenta] would be a problem, since we would be hard pressed to take the derivatives we need in order to compute the canonical transformation [...] To proceed, we take advantage of the fact that because of the analyticity of H, the Fourier coefficients $h_k(\hat{p})$ are decaying to zero exponentially fast as $|k|$ becomes large. Thus, if we truncate the sum defining S to consider only $|k| < N$, for some large N we will make only a relatively small error in the solution [...]. On the other hand, since there are now only finitely many terms in the sum defining S, we can find open sets of action-variables on which the generating function is defined. (Wayne 1996, p. 18; symbols adapted)

Especially, when the frequency changes due to $H_{s-1}^+(p_{s-1}) \to H_s^+(p_s)$ in the s-th canonical transformation, one can choose δ_s of the transformation sufficiently small that the new frequency obeys a (somewhat weaker) Diophantine condition if the old frequency does. Indeed, let

$$|(k, \omega)| \geq \frac{K}{|k|^{n+1}}. \tag{3.77}$$

Then, dropping the iteration subscript s, one has for $\omega = \partial H^+/\partial p|_p$ and $\omega' = \partial H^+/\partial p|_{p'}$ (the following corresponds to Eq. 107 in Chierchia 2009, p. 5077; compare also with Eq. 6, p. 21, in Pöschel 2001 and the last equation on p. 19 of Wayne 1996),

$$
\begin{aligned}
|(k,\omega')| &= \left| (k,\omega) + \left(k, \left.\frac{\partial H^+}{\partial p}\right|_{p'} - \left.\frac{\partial H^+}{\partial p}\right|_p \right) \right| \\
&\overset{(a)}{=} \left| (k,\omega) + \left(k, \left.\frac{\partial^2 H^+}{\partial p^2}\right|_{\bar{p}} (p'-p) \right) \right| \\
&\overset{(b)}{\geq} |(k,\omega)| \left(1 - \frac{\|\partial^2 H^+/\partial p^2\|}{|(k,\omega)|} |k|\,|p'-p| \right) \\
&\overset{(c)}{\geq} \frac{K}{|k|^{n+1}} \left(1 - \frac{n\Theta}{K} |k|^{n+2} |p'-p| \right) \\
&\overset{(d)}{\geq} \frac{K}{|k|^{n+1}} \left(1 - \frac{n\Theta N^{n+2}}{K} |p'-p| \right),
\end{aligned}
\tag{3.78}
$$

where (a) applies the mean value theorem with some intermediate $\bar{p}$, (b) uses $\big| |a| - |b| \big| \leq |a+b|$ and the supremum norm, (c) is the Diophantine condition and uses $\|\partial^2 H^+/\partial p^2\| \leq n\Theta$ for given $1 < \Theta < \infty$ from Eq. (4.81); and finally (d) is the N-cutoff. Since $|p'-p| < \delta^3$ from Eq. (4.77), for small δ the new divisor $|(k,\omega')|$ in Eq. (3.78) obeys almost the same Diophantine condition as $|(k,\omega)|$.

> [T]he remainder term R_n can be regarded as part of the 'terms of higher order' ... This process was already used by Bogolubov, Mitropolski [in 1962]. (Arnold & Avez 1968, p. 255)

And:

> It is not difficult to see that if, for $|\mathrm{Im}\, q| \leq \rho$, we have $|f(q)| \leq M$, then the remainder of the Fourier series admits an estimate of the form $|R_N f(q)| \leq M^2$ for $|\mathrm{Im}\, q| \leq \rho - \gamma$ if $N = \frac{1}{\gamma} \ln \frac{1}{M}$. (Arnold 1963b, p. 105; symbols adapted)

Arnold thus introduces functions $[H]_N$ and $[S]_N$ that are trigonometric polynomials of degree $< N$,

$$
[H(p,q)]_N = \sum_{0<|k|<N} h_k(p)\, e^{i(k,q)}.
\tag{3.79}
$$

Sternberg (1969) introduces a linear operator T_N (named S_N in eq. 13.7, p. 135) such that, avoiding possibly misleading rectangular brackets for the argument H,

$$
H_N = T_N(H),
\tag{3.80}
$$

and follows Nash's electrical terminology to call T_N a 'low pass filter' since 'it does not disturb low frequencies.' Sternberg (1969), p. 144, also refers to T_N as a 'smoothing operator:'

> The first naive idea that occurs to one when trying to approximate a function by a smoother one is just to expand in Fourier series and to keep only a finite number of terms corresponding to the harmonics of small degree. (de la Llave 2001, p. 28)

However,

> keeping only a finite number of the low order Fourier terms is a much less efficient method of approximation (from the point of view of the number of derivatives required) than convolving with a smooth kernel. (de la Llave 2001, p. 28)

Dumas (2014), p. 72, names $[H]_N$ an 'ultraviolet cutoff,' a term borrowed from quantum field theory.

3.7 DOMAINS WITHOUT INTERIOR

> The above considerations do not exclude the possibility that invariant tori on which $(\omega, k) \neq 0$ may exist in a perturbed system. These tori cannot, however, fill any domain. (Arnold 1963b, p. 98)

Consider an integrable Hamiltonian system with n degrees of freedom and arbitrary frequency vector $(\omega_i) \in \mathbb{R}^n$ (for motion on a phase-space torus). Then (ω_i) can be approximated to any precision by a rational vector $(w_i) = (a_i/b_i) \in \mathbb{Q}^n$, and the w_i are in resonance (i.e., not linearly independent over $\mathbb{Z}$). For example, if n is even, then $\sum_{i=1}^n k_i w_i = 0$ with $k_i \in \mathbb{Z}$ is obtained for

$$k_i = \begin{cases} b_i a_{n+1-i}/a_i & \text{if } i \leq n/2 \\ -b_i & \text{if } i > n/2. \end{cases} \tag{3.81}$$

Thus frequency vectors with resonances lie dense in frequency space, and if all resonance frequencies are removed for $s \to \infty$ (sequence of canonical transformations), a set with unique topological properties remains: it has no interior points, since any neighborhood of a point $p \in G$ contains points that lie not in G. Similarly, the preimage G in momentum space of non-resonant frequencies under the diffeomorphism A has a dense set of points excluded. Thus $H(p, q)$ is defined on a set G (times $(2\pi)^n$) without interior points.

A function $f(z)$ defined on a set E of the complex plane $\mathbb{C}$ is called *monogenic* (with respect to E) at a finite non-isolated point $\zeta \in E$ if it has a finite derivative

$f'_E(\zeta)$ with respect to $z \in E$ at this point:

$$f'_E(\zeta) = \lim_{\substack{z \to \zeta \\ z \in E}} \frac{f(z) - f(\zeta)}{z - \zeta}.$$

A function which is monogenic at every non-isolated point of E is called monogenic on E. (Dolzhenko 1995, p. 7)

With the cutoff N, one avoids that resonances form a dense set, and can work with analytic instead of monogenic functions. For $s \to \infty$ and $N_s \to \infty$, eventually all resonances are included and frequency and momentum are defined on sets without interior – but all required derivatives were formed on domains with interior points during the s-iteration. It will be shown in the next chapter that for vanishing perturbations, the measure of the 'porous' sets G and Ω containing the invariant tori approaches the full measure of the momentum and frequency domains.

We summarize some basic facts from topology and measure theory about *perfect, nowhere dense* sets of *finite measure*. It is assumed that the *open sets* of the topological space X are already defined. All sets are assumed to be subsets of X. A *neighborhood* U_p of a point $p \in X$ is a set so that there is an open set O with $p \in O \subset U_p$. Let $A \subset X$ be an arbitrary subset.

A point $p \in A$ is an *interior point* of A if there is an open set O with $p \in O \subset A$. Thus A is a neighborhood of p. The *interior* $I(A)$ of A is the set of all interior points of A. Equivalently, $I(A)$ is the union of all open sets contained in A. For open sets, $A = I(A)$.

A point p is *limit point* or *accumulation point* of A if any neighborhood U_p contains a point $q \in A$ and $q \neq p$. Thus 0 is a limit point of $A = \{1/n : n \in \mathbb{N}\}$ but $0 \notin A$. The *closure* $C(A)$ of A is the union of A and all its limit points. A is *closed* if $A = C(A)$. The *derived set* A^d of A is the set of all limit points of A. If A is closed, it contains all its limit points, $A^d \subset A$. Conversely, if $A^d \subset A$, A contains its limit points and thus is closed. Therefore $A = C(A) \leftrightarrow A^d \subset A$.

A point $p \in A$ is an *isolated point* of A if there is a neighborhood U_p that contains no point of A except p. $p \in A$ is isolated $\leftrightarrow p \in A \backslash A^d$ (Kuratowski 1966, p. 75). A set is *discrete* if each of its points is isolated.

The set A is *dense in itself* if no point of A is an isolated point, i.e., $p \notin A \backslash A^d$ for any $p \in A$. This is the same as $p \in A \to p \in A^d$, or $A \subset A^d$. A is *perfect* if $A = C(A)$ and A is dense in itself, i.e., A is closed and has no isolated points.

Lemma (see Kuratowski 1966, p. 77) *A is perfect $\leftrightarrow A = A^d$.*

Proof A is perfect $\leftrightarrow A = C(A)$ and A is dense in itself $\leftrightarrow A^d \subset A$ and $A \subset A^d$. $\square$

The set A is *dense* in X if any neighborhood of any point of X (!) contains a point of A. Then $C(A) = X$.

A subset $A \subset X$ is *nowhere dense* in X if $I(C(A)) = \emptyset$, i.e., $C(A)$ contains no open subset except $\emptyset$.

'Nowhere dense' is *not* the opposite (logical negation) of 'dense.' For example, the interval $[0,1]$ is not dense in $\mathbb{R}$ (there are neighborhoods of 2 that do not meet $[0,1]$), but $[0,1]$ is also not nowhere dense, for $\text{int}(\text{clos}([0,1])) = \text{int}([0,1]) = (0,1) \neq \emptyset$. Still, there are sets that are both 'not dense' and 'nowhere dense,' like $\mathbb{Z} \subset \mathbb{R}$: 'not dense' is obvious; as for 'nowhere dense,' $\text{int}(\text{clos}(\mathbb{Z})) = \text{int}(\mathbb{Z}) = \emptyset$.

A *meagre* set is a countable union of nowhere dense sets. The rationals $\mathbb{Q}$ are meagre in $\mathbb{R}$. Nowhere dense sets and thus meagre sets can be uncountable: $\mathbb{R}$ is nowhere dense in $\mathbb{R}^2$, since, with respect to the topology of $\mathbb{R}^2$, $\text{int}(\text{clos}(\mathbb{R})) = \text{int}(\mathbb{R}) = \emptyset$.

An example of a perfect and nowhere dense set is given by the Cantor set T, obtained by removing from the real interval $[0,1]$ the open middle third $(\frac{1}{3}, \frac{2}{3})$, from the remaining two closed intervals again the open middle thirds, from the remaining four closed intervals the open middle thirds, etc. T is the set of all $c \in \mathbb{R}$ that have ternary expansion (basis 3; digits 0,1,2) missing the digit 1,

$$ c = \sum_{i=1}^{\infty} c_i 3^{-i}, \qquad c_i \in \{0,2\}. \tag{3.82} $$

As intersection of closed sets, T is closed. Let $\varepsilon' > 0$ be arbitrary and $N > 1$ so that $\varepsilon = 3^{-N} < \varepsilon'$. Then all d that agree with c in the first N ternary digits and have no digit 1 afterward belong to T and have distance $< \varepsilon'$ from c. Thus each $c \in T$ is a limit point, and T is perfect. Assume that $c \in T$ is an interior point. Then there is a neighborhood $U_c \subset T$, where $U_c = I$ for some interval $I \subset \mathbb{R}$. Assume $d \in U_c$ and $c \neq d$, and that c and d agree in the first $N-1$ ternary digits, but not in the N-th. Then the N-th digits of c and d are either 0 and 2, respectively, or 2 and 0. The number that agrees with c and d in the first $N-1$ digits and has 1 as N-th digit lies between c and d but does not belong to T. Contradiction to $U_c \subset T$, thus T has no interior points and is nowhere dense. The Cantor set has measure zero:

$$ \text{mes}\,(T) = 1 - \frac{1}{3} - \frac{2}{3^2} - \frac{4}{3^3} - \cdots = 1 - \frac{1}{3}\sum_{n=0}^{\infty}\left(\frac{2}{3}\right)^n = 1 - \frac{1/3}{1 - 2/3} = 0. \tag{3.83} $$

There exist variants of the Cantor set, called Smith-Volterra-Cantor sets (introduced even before the Cantor set), which are nowhere dense but have finite measure. For example, remove the open middle $1/4$ of the interval $[0,1]$, of the remaining two closed intervals remove the middle open $1/4^2$, of the remaining four closed intervals remove the open middle $1/4^3$, and so forth. Although the removed intervals become small very quickly, each interval in $[0,1]$ suffers removals, thus there are no full intervals left in the resulting set, which has, therefore, empty interior. The removed intervals have length/measure of

$$ \frac{1}{4} + \frac{2}{4^2} + \frac{4}{4^3} + \frac{8}{4^4} + \cdots = \frac{1}{4} + \frac{1}{8} + \frac{1}{16} + \frac{1}{32} + \cdots = \frac{1}{2}, \tag{3.84} $$

and therefore the Smith-Volterra-Cantor set has also measure $1/2$. Another example is the following:

[E]numerating the rationals in $[0,1]$ as q_n and choosing an open interval I_n of length $1/3^n$ containing q_n for each n, then the union of these intervals has measure at most $1/2$. Hence, the set of points in $[0,1]$ but not in any of I_n has measure at least $1/2$, despite being nowhere dense. (Wolfram MathWorld, entry 'Nowhere Dense')

In the KAM theorem, the invariant tori of the perturbed Hamiltonian form a perfect, nowhere dense set of finite measure: the Diophantine condition and boundary removals leave closed sets of frequencies and momenta. Each neighborhood of each p corresponding to an invariant torus contains $p' \neq p$ of another invariant torus, thus p is an accumulation point; but it also contains p'' that corresponds to a resonance, thus *not* to an invariant torus, hence p is not an interior point. The finite measure will be demonstrated in the next chapter. We also note the following important fact, but do not go further into this.

In KAM theory, we often have to study functions defined in Cantor sets. In particular, sets with empty interior. In this situation, the concept of Whitney differentiability plays an important role. A reasonable notion of smooth functions in closed sets is that they are the restriction of smooth functions in open sets that contain them. This definition is somewhat unsatisfactory since the extension is not unique. In the paper Whitney (1934), one can find an intrinsic characterization of smooth functions in a closed set. (de la Llave 2001, p. 30)

3.8   ANALYTIC FUNCTIONS

The functions A, B, H, S (frequency map; canonical diffeomorphism; Hamilton function; generating function of canonical transformation) are assumed to be analytic in complex domains (complex analytic, or simply analytic). Actually, real analytic functions would suffice, since a standard theorem gives then complex analyticity in a finite strip:

Theorem *Let $f : \mathbb{R}^n \to \mathbb{R}$ be a real analytic function with period 2π in all variables. Then there exists an $r > 0$ and a complex analytic function $g : G \to \mathbb{C}$ with $g|_{\mathbb{R}^n} = f$ (that is, f and g agree for real arguments). Here G is the set of all complex n-tuples $(z_1, \ldots, z_n)$ with $|\mathrm{Im}\, z_i| < r$ for all i, and g is 2π-periodic with respect to all $\mathrm{Re}\, z_i$.*

For the *proof*, see Rüssmann (1979), p. 178. The definition of a *real analytic function* is as follows:

A function f, with domain an open set $U \subset \mathbb{R}$ and range either the real or the complex numbers, is said to be *real analytic* at α if the function f may be represented by a convergent power series on some interval of positive radius centered at α:

$$f(x) = \sum_{j=0}^{\infty} a_j (x - \alpha)^j. \tag{3.85}$$

The function is said to be *real analytic* on $V \subset U$ if it is real analytic at each $\alpha \in V$. (Krantz & Parks 2002, p. 3)

For the definition of a real analytic function $f : U \subset \mathbb{R}$ with U an open subset of $\mathbb{R}^m$, see Krantz & Parks (2002), p. 29. Real and complex analytic functions are infinitely often differentiable. But whereas complex functions that are (once) complex differentiable are analytic, real functions can be infinitely differentiable but *not* real analytic, if their Taylor series does not converge. For the proof, see Markushevich (1965), pp. 6-7.

We assume here, as does Arnold (1963a,b), that all functions are complex analytic in a complex strip of finite width. Actually, if one thinks of a 2π-periodic function of one argument φ to be defined on the circumference of a circle with azimuth φ as coordinate, one should replace 'strip' by 'annulus.'

The importance of assuming complex analytic functions becomes clear from the frequent reference to Im q and Im Q in the next chapter. In the first place, Cauchy's integral theorem $\oint dz\, f(z) = 0$ is available then, and allows estimates for Fourier coefficients and for derivatives of functions. This 'Cauchy estimate [...] is used over and over' (Pöschel 2001, p. 28). The required estimates are obtained in Lemmas A2 and A3 of Chapter 5. Moser (1962), using the smoothing technique introduced by Nash, proved a version of the KAM theorem for functions that are not (real) analytic, but can be differentiated at least 333 times.

3.9 MOTION ON A TORUS

Motion on a phase-space torus for integrable systems was discussed in Sections 1.6, 1.7 and 1.8. We note here the rather obvious fact that the angle coordinates q_i, $i = 1,\dots,n$ on the torus can be thought of as straight Cartesian coordinates either in full $\mathbb{R}^n$ with $-\infty < q_i < \infty$, or along the edges of a cube with $0 \le q_i \le 2\pi$; some texts speak then of unspooled or unwound coordinates. If the ω_i are linearly independent over $\mathbb{Z}$, i.e., there are no resonances among them, both the torus and the cube are filled uniformly by the trajectory, which passes through every neighborhood of every point. This leads to an ergodic theorem (equality of time averages and phase-space averages) first proved by H. Weyl, see Arnold et al. (1997), p. 111, or Chierchia (2009), p. 5067. The KAM theorem states that after an infinite number of infinitesimal canonical transformations of the perturbed Hamiltonian and for appropriate initial conditions p_0, the motion

$p_\infty(t), q_\infty(t)$ is a quasi-periodic solution of Hamilton's equations [...] [T]his argument is independent of the point q_0 that we take on the original torus. Thus it shows that *every* trajectory on the unperturbed torus is preserved. (Wayne 1996, p. 26; symbols adapted)

3.10 SMALL NUMBERS

[L]imit processes play a role in every analytic problem. (Rudin 1991, p. 3)

The roles of ε and δ in standard calculus are adopted by $\delta, M, \beta, \gamma, \kappa, K$ in Arnold's proof, with κ and M resembling ε, and δ, β, γ, K resembling δ. For the following discussion see Arnold (1963b), p. 106, and Wayne (1996), pp. 23-24.

κ with $0 < \kappa < 1$ is an arbitrary, fixed number and gives the relative measure of the set of invariant tori that survive the perturbation. Obviously, $\kappa \to 1$ requires vanishing perturbation, $M_1 \to 0$.

M_s with $s = 1, 2, \ldots$ is the upper bound of the perturbation of the Hamiltonian, $\|H_s^-\| \le M_{s+1}$. The initial perturbation of the integrable Hamiltonian is limited to $\|H_0^-\| \le M_1(\kappa)$. An iteration sequence is set up for M_s, with $M_s \to 0$ for $s \to \infty$ (the same is true for $\delta_s, \beta_s, \gamma_s$). M_s is given as a rather large power of δ_s, $M_s = \delta_s^T$ with $T = 8n + 24 \ge 40$ (the case $n = 1$ is irrelevant). At iteration start, δ_1 and thus M_1 have to be sufficiently small in order to obtain convergence, $\delta_1 < 1/20$ or less (see Eq. 4.108, where $\Theta_{s-1} > 1$). Therefore, $M_1 < 20^{-40} \approx 10^{-52}$ for the allowed size of perturbations of the integrable Hamiltonian. In terms of the natural constants of physics, this is very small – mathematically, however, it is finite.[9] Arnold (1963a,b) makes no attempt to obtain 'optimal' bounds. In numerical solutions of (chaotic) non-linear systems, one finds 'tori' (usually not phase-space tori in the narrow sense) that survive perturbations of order unity.

δ_s is the most important number and controls all upper bounds. Therefore, it appears ultimately in all bounds. The iteration sequence of the Newton-like scheme of canonical transformations uses $\delta_{s+1} = \delta_s^{3/2}$.

$\beta_s = \delta_s^3$ controls the size of both the frequency domain Ω_s and the momentum domain G_s. In the s-th iteration step, Ω_s is given by $\Omega_s = \Omega_{s-1} - (5 + 7\Theta_{s-1})\beta_s$, apart from a reduction by the Diophantine condition. Thus β_s must fall off sufficiently fast to leave a non-empty set Ω_∞. The momentum domain G_s is obtained from the inverse A_s^{-1} of the frequency diffeomorphism A_s. For a fixed frequency map (no perturbations of frequencies), the reduction of G_s would be $G_s = G_{s-1} - 2\beta_s$ only. β_s is also the measure for the deviation of the canonical transformation B_s from the identity map E.

$\gamma_s = \delta_s^{1/(4n)}$ controls the reduction of the width ρ_s of the complex strip of the angle variable q_s (in order that Cauchy's theorem is available for Cauchy estimates). As is the case for β_s, γ_s must fall off sufficiently fast to leave a strip of finite width ρ_∞ for q_∞. Moreover, the cutoff N_s of Fourier series is proportional to γ_s^{-1}. This makes γ_s a somewhat sensitive choice: γ_s has to be small enough to leave a finite complex strip for q_∞, but large enough to avoid too quick inclusion of 'all' resonances as $N_s \to \infty$.

---

[9] In contrast to 'linear perturbation theory' in physics, which addresses stability in the limit of vanishing perturbations only, the KAM theorem states stability of a (large) set of phase-space tori under small yet finite perturbations. It is now known from numerical experiments that small regions of nested tori can exist within large, chaotic domains up to perturbation amplitudes of order unity.

$K = \delta_1 \kappa$ is the proportionality constant (not iterated!) in the Diophantine condition. It thus controls the measure of the set of surviving non-resonant tori.

The decay rate of δ_s, and along with it, of β_s and γ_s, have to be chosen as a compromise:

> [I]f we choose δ_s going to zero slowly, we lose more domain than needed [...] – of course, if we lose too fast, we end up with an empty domain. On the other hand, the smaller that we choose δ_s, the worse [3.75] becomes. A reasonable compromise that is neither too fast so that we end up with no domain nor too slow so that we can still converge is to choose an exponential rate of decay. (de la Llave 2001, p. 66; our equation number)

To obtain superfast convergence $M_{s+1} = M_s^\alpha$ with $\alpha > 1$, specifically $\alpha = 3/2$, Arnold (1963a) chooses $T = 8n + 24$. As discussed above, this large T is the reason why the allowed perturbations are very small. To summarize:

> These assumptions, that we had to consider in the proofs based on composition [i.e., the infinite sequence of canonical transformations] are subtle because they require that the errors decrease faster than the analyticity losses. (de la Llave 2001, p. 115)

The total measure of the lost resonance strips scales with N_s^n, thus to keep this loss small, N_s has to be sufficiently small and γ_s sufficiently large; Arnold chooses $\gamma_s = \delta_s^{1/(4n)}$ (for this discussion, see items 1 to 3 in Arnold 1963b, p. 106). Obviously, the start value for δ_1 depends on the given constants of the problem (β_1 and γ_1 have to be brought to the appropriate scales), which are $n, D, \kappa, \rho_0, \theta_0, \Theta_0$, see Eqs. (4.107) and (4.108). The constants $D, \rho_0, \theta_0, \Theta_0$ are introduced in the next chapter. Altogether, one has for the small numbers (see Arnold 1963b, p. 105) that

$$M_s \ll \beta_s \ll \delta_s \ll \gamma_s. \tag{3.86}$$

4 Proof of the KAM Theorem

This chapter gives the full proof of the KAM theorem. It consists of three theorems – the Fundamental theorem, the Inductive theorem and the KAM theorem. These theorems rely on 20 technical lemmas proved in the following chapters.

The Fundamental theorem merges the Fundamental lemma and the Inductive lemma of Arnold (1963a). It addresses the frequency shift from the integrable part $\bar{H}$ of the perturbation (see Eq. 3.57) and the infinitesimal canonical transformation to higher perturbation order, $M_s \to M_{s+1}$. The small divisor problem in the latter is overcome using a Diophantine condition that excludes resonances.

The Inductive theorem sets up an infinite sequence of frequency shifts and infinitesimal canonical transformations via a control parameter $\delta_s \to \delta_{s+1} = \delta_s^{3/2}$ (see de la Llave 2001, p. 96, item F.1). This sequence converges to an integrable Hamiltonian on a momentum set with finite measure.

Finally, the KAM theorem relates the slightly deformed phase-space tori of the perturbed system to the tori of the unperturbed integrable system.

An infinite sequence of Hamilton functions $H_s(p_s, q_s)$ is introduced, where $H_s(p_s, q_s) = H_s^+(p_s) + H_s^-(p_s, q_s)$, with integrable part $H_s^+(p_s)$ and small perturbation $H_s^-(p_s, q_s)$. For a single step $(s-1) \to s$ in the sequence of canonical transformations, $H_{s-1}(p_{s-1}, q_{s-1}) = H_s(p_s, q_s)$, since the generating function S is independent of time. Put $(p, q) = (p_{s-1}, q_{s-1})$ and $(P, Q) = (p_s, q_s)$. Then the transformation of the Hamiltonian takes the following form.[1]

$$
\begin{aligned}
 & H_{s-1}^+(p) && + && H_{s-1}^-(p, q) \\
= \;\; & \underbrace{H_{s-1}^+(p) \;\; + \;\; \overline{H}(p)} && + && \overbrace{\tilde{H}(p, q)} \\
= \;\; & H_s^+(p) && + && \tilde{H}(p, q) \\
= \;\; & H_s^+(P) && + && H_s^-(P, Q) && (4.1)
\end{aligned}
$$

Notice the transition from $H_s^+(p)$ to $H_s^+(P)$ in the last line, which is discussed in Eq. (4.45).

---

[1] With the diffeomorphism $B_s(x_s) = x_{s-1}$ one has $H_s(x_s) = H_{s-1}(x_{s-1}) = H_{s-1}(B_s(x_s))$ and thus $H_s = H_{s-1} \circ B_s$, see Eq. (82) in Chierchia (2009).

DOI: 10.1201/9781003287803-4

4.1 FUNDAMENTAL THEOREM

Fundamental theorem[2] *Consider the functions* $H_{s-1}(p_{s-1}, q_{s-1}) = H_{s-1}^{+}(p_{s-1}) +$
$H_{s-1}^{-}(p_{s-1}, q_{s-1})$ *and* $A_{s-1}(p_{s-1})$ *for a Hamilton system with n degrees of free-*
dom, the compact complex domains F_{s-1}, G_{s-1}, Ω_{s-1} *and positive real numbers*
$K, \rho_{s-1}, \theta_{s-1}, \Theta_{s-1}, \delta_s, \beta_s, \gamma_s, M_s$. *Assume that:*

(A) *Domain and bound of A_{s-1}*

The function [3] A_{s-1} *defined by*

$$A_{s-1}(p_{s-1}) = \frac{\partial H_{s-1}^{+}(p_{s-1})}{\partial p_{s-1}} \tag{4.2}$$

is a diffeomorphism from G_{s-1} *onto* Ω_{s-1}, *with*

$$\theta_{s-1} |dp_{s-1}| \leq \|dA_{s-1}\| \leq \Theta_{s-1} |dp_{s-1}|, \tag{4.3}$$

where $0 < \theta_{s-1} < 1 < \Theta_{s-1} < \infty$.

(B) *Domain and bound of H_{s-1}*

The function $H_{s-1}(p_{s-1}, q_{s-1})$ *is analytic on the domain*

$$F_{s-1} : \quad p_{s-1} \in G_{s-1}, \qquad |\mathrm{Im}\, q_{s-1}| \leq \rho_{s-1}, \tag{4.4}$$

where $\rho_{s-1} \leq 1$ *and* $\mathrm{Re}\, q_{s-1}$ *is 2π-periodic in each component. On* F_{s-1},

$$\left\| H_{s-1}^{-} \right\| \leq M_s. \tag{4.5}$$

(C) *Smallness of $\delta_s, \beta_s, \gamma_s, M_s$*

The small numbers $\delta_s, \beta_s, \gamma_s, M_s, K$ *satisfy* [4]

$$\delta_s < \delta^{(1)}(n, \theta_{s-1}, \Theta_{s-1}) = \min\left\{ \frac{1}{L_1}, \frac{1}{L_2}, \frac{1}{2\,\Theta_{s-1} L_3}, \frac{1}{L_4}, \frac{\theta_{s-1}}{2n} \right\}, \tag{4.6}$$

$$15\delta_s < 3\gamma_s < \rho_{s-1} \leq 1, \tag{4.7}$$

$$3\beta_s < 2\delta_s, \tag{4.8}$$

$$2\beta_s < K \leq 1, \tag{4.9}$$

$$M_s < \delta_s^{\nu} K \beta_s^2, \tag{4.10}$$

---

[2]Lemma G5 in Section 6.9 on the perturbed frequency diffeomorphism plays a crucial role in the Fundamental theorem and should be inserted here.

[3]For functions of several complex variables, see Section 5.2. Furthermore, '[w]e remark that, in what follows, the analyticity domains of actions and angles play a different role' (Chierchia 2009, p. 5075).

[4]Note that $M_s \ll K, \delta_s, \beta_s, \gamma_s$. In Eq. (4.3.5) of Arnold (1963b), the restriction on γ is given correctly as $3\gamma \leq \rho$, whereas Arnold (1963a), p. 17, has still $2\gamma \leq \rho$, as in his Fundamental lemma.

where $v = 2n+3$ and

$$L_1 = 16,$$
$$L_2 = 4^{n+2} n e^{-(n+1)} (n+1)^{n+1},$$
$$L_3 = 2^{4n-1} n^2 e^{-(2n+2)} (n+1)^{2n+2},$$
$$L_4 = 2^{n+1} e^{-n} n^n. \tag{4.11}$$

Then the following holds:

(I) *Domain and bound of A_s*

The function A_s defined by

$$A_s(p_s) = \frac{\partial H_s^+(p_s)}{\partial p_s} = \omega \tag{4.12}$$

is a diffeomorphism from G_s onto Ω_s. Here [5]

$$G_s = A_s^{-1}(\Omega_s), \tag{4.13}$$
$$\Omega_s = (\Omega_{s-1})_{KN_s} - d_s, \tag{4.14}$$
$$(\Omega_{s-1})_{KN_s} = \left\{ \omega \in \Omega_{s-1} : \; |(k,\omega)| \geq \frac{K}{|k|^{n+1}} \right\} \tag{4.15}$$

with

$$N_s = \frac{1}{\gamma_s} \ln \frac{1}{2M_s}, \tag{4.16}$$
$$d_s = (5+7\Theta_{s-1})\beta_s \tag{4.17}$$

and [6] $0 < |k| = |k_1| + \cdots + |k_n| < N_s$ *for $k \in \mathbb{Z}^n$. Furthermore,* [7]

$$\theta_s |dp_s| < \|dA_s\| < \Theta_s |dp_s|, \tag{4.18}$$

where

$$\theta_s = (1-\delta_s)\,\theta_{s-1}, \qquad\qquad \Theta_s = (1+\delta_s)\,\Theta_{s-1}. \tag{4.19}$$

---

[5] A_{s-1} corresponds to A in Lemma G5, and A_s to the perturbed A'. Thus Eq. (6.98) becomes here $G_s = A_s^{-1}(\Omega_s)$.

[6] Higher resonances are still allowed in this iteration step s.

[7] Lemma G5 has $|dA| \leq \Theta |dp|$ and $|dA'| < (1+\delta)\Theta|dp|$ and $d = (5+7\Theta)\beta$ for $b = 3\beta$. This becomes here $|dA_{s-1}| \leq \Theta_{s-1}|dp|$ and $|dA_s| < (1+\delta_s)\Theta_{s-1}|dp|$ and $d_s = (5+7\Theta_{s-1})\beta_s$ for $b = 3\beta_s$. Arnold (1963a) replaces the latter by $d_s = (5+7\Theta_s)\beta_s$, apparently for easier readability. This is allowed since $\Theta_{s-1} < \Theta_s$, and d_s is the width of the strip taken off the frequency domain; thus Arnold allows for a larger frequency loss.

The change in A is

$$\|A_s - A_{s-1}\| < \beta_s \,\delta_s. \tag{4.20}$$

(II) *Measure of lost domain*

Let $\overline{d}_s = d_s + \beta_s$ and [8] $\overline{\Omega}_s = (\Omega_{s-1})_{KN_s} - \overline{d}_s$. *Then*

$$\mathrm{mes}\,(G_{s-1} \backslash G_s) \leq \frac{1}{\theta_{s-1}^n}\, \mathrm{mes}\,\left(\Omega_{s-1} \backslash \overline{\Omega}_s\right). \tag{4.21}$$

(III) *Domain and bound of B_s*

There is a diffeomorphism $B_s : F_s \rightarrow F_{s-1}$, $B_s(p_s, q_s) = (p_{s-1}, q_{s-1})$ *on*

$$F_s: \qquad p_s \in G_s, \qquad |\mathrm{Im}\, q_s| \leq \rho_{s-1} - 3\gamma_s, \tag{4.22}$$

$F_s \subset F_{s-1} - \beta_s$, *and* B_s *obeys the bounds* $(x_s = (p_s, q_s))$

$$\|B_s - E\| < \beta_s, \qquad \|dB_s\| < 2\,|dx_s|. \tag{4.23}$$

(IV) *Domain and bound of H_s*

For $(p_s, q_s) \in F_s$ *one has* [9]

$$H_{s-1}(p_{s-1}, q_{s-1}) = H_s^+(p_s) + H_s^-(p_s, q_s) \tag{4.24}$$

with [10]

$$\|H_s^-\| < M_{s+1}, \tag{4.25}$$
$$\|\partial H_s^- / \partial x_s\| < M_{s+1}/\beta_s, \tag{4.26}$$
$$\|\partial^2 H_s^- / \partial x_s^2\| < 2M_{s+1}/\beta_s^2, \tag{4.27}$$

where

$$M_{s+1} = M_s^2\, \delta_s^{-2\nu}\, \beta_s^{-2}. \tag{4.28}$$

Remark Kolmogorov originally suggested $\|H_s^-\| < M_s^2$. In Eq. (4.28), this is reduced by the factor $\delta_s^{-2\nu} \beta_s^{-2} \gg 1$ to $\|H_s^-\| < M_s^{3/2}$. Any power larger than 1 works;

---

[8] Similar to Θ_s in Footnote 7, one can use θ_s in Eq. (4.21). One has $1/\theta_{s-1}^n < 1/\theta_s^n$, so the upper bound for the lost domain becomes weaker when using θ_s instead of θ_{s-1}.

[9] There is a double misprint $H(p)$ instead of $\overline{H}(P)$ in subitem 2 of item 3° of Lemma 2.3 in Arnold (1963a), and a misprint $\overline{H}(p)$ in the Russian original. In Arnold (1963b), the corresponding Eq. (4.3.7) is correct.

[10] Arnold mostly has $\|H_s^-\| \leq M_{s+1}$, but we retain his use of $\|H_s^-\| < M_{s+1}$ here (see item 2. on p. 17 in Arnold 1963a).

Arnold (1963b), Eq. (4.2.9) has $\|H_s^-\| < M_s^{15/14}$. The number $T \in \mathbb{N}$ is chosen so that the perturbation series convergences: 'We choose T to be sufficiently large so that $[M_{s+1} = M_s^2 \beta_s^{-2} \delta_s^{-2\nu} < M_s^{3/2} \ldots]$ the approximations then converge' (Arnold 1963b, p. 106).

Proof Statements (I) and (II) on A_s, G_s, Ω_s are proved using Lemma G5 (frequency variation lemma). Let

$$(p,q) = (p_{s-1}, q_{s-1}), \qquad (P,Q) = (p_s, q_s), \qquad (4.29)$$

and put

$$H_{s-1}^-(p,q) = \overline{H}(p) + \tilde{H}(p,q), \qquad (4.30)$$

where

$$\overline{H}(p) = \frac{1}{(2\pi)^n} \int dq \, H_{s-1}^-(p,q), \qquad (4.31)$$

and thus

$$\int dq \, \tilde{H}(p,q) = 0 \qquad (4.32)$$

and

$$\begin{aligned} H_{s-1}(p,q) &= H_{s-1}^+(p) + \overline{H}(p) + \tilde{H}(p,q) \\ &= \quad H_s^+(p) \quad + \tilde{H}(p,q). \end{aligned} \qquad (4.33)$$

Let

$$A_s(p) = \frac{\partial H_s^+}{\partial p} = \frac{\partial H_{s-1}^+}{\partial p} + \frac{\partial \overline{H}}{\partial p} = A_{s-1}(p) + \Delta(p). \qquad (4.34)$$

From Eq. (4.5), $|H_{s-1}^-(p)| \le M_s$ for $p \in G_{s-1}$, and from Eq. (4.31), also $|\overline{H}(p)| \le M_s$. A rough estimate gives

$$M_s \overset{(a)}{<} \delta_s^{2n+3} K \beta_s^2 \overset{(b)}{\le} \delta_s^{2n+3} \beta_s^2 \overset{(c)}{<} \delta_s^2 \beta_s^2 \overset{(d)}{<} \frac{\delta_s \theta_{s-1}}{2n} \beta_s^2, \qquad (4.35)$$

using Eq. (4.10) in (a), $K \le 1$ in (b), $\delta_s < 1$ in (c) and $\delta_s < \theta_{s-1}/(2n)$ from Eq. (4.6) in (d). The Cauchy estimate (5.33) from Lemma A3 gives, for $p \in G_{s-1} - \beta_s$,

$$|\Delta(p)| = \left| \frac{\partial \overline{H}(p)}{\partial p} \right| \le \frac{M_s}{\beta_s} < \frac{\delta_s \theta_{s-1}}{2n} \beta_s < \beta_s \delta_s, \qquad (4.36)$$

since $\theta_{s-1} < 1$. This is Eq. (4.20), since $G_s \subset G_{s-1} - \beta_s$ (see below). Furthermore, still for $p \in G_{s-1} - \beta_s$,

$$
\begin{aligned}
|d\Delta(p)| = \left| d\,\frac{\partial \overline{H}(p)}{\partial p} \right| &= \max_i \left(\left| \sum_{j=1}^{n} \frac{\partial^2 \overline{H}}{\partial p_i \partial p_j}\, dp_j \right| \right) \\
&\leq \max_i \left(\left| \sum_{j=1}^{n} \frac{\partial^2 \overline{H}}{\partial p_i \partial p_j} \right| \right) \max_j (|dp_j|) \\
&\leq n \max_{i,j} \left(\left| \frac{\partial^2 \overline{H}}{\partial p_i \partial p_j} \right| \right) |dp| \\
&\leq n\,\frac{2M_s}{\beta_s^2}\, |dp| < n\,\frac{\delta_s \theta_{s-1}}{n}\, |dp| = \delta_s \theta_{s-1}\, |dp|,
\end{aligned}
\qquad (4.37)
$$

using Eq. (5.33) for the second derivatives. Identifying $A \equiv A_{s-1}$ and $A' \equiv A_s$, all assumptions of Lemma G5 are met, especially the bounds in Eqs. (6.70) to (6.72). The number $b > 0$ and the set[11] Ω_0 are chosen so that, for

$$
(G_s)_{KN_s} = A_s^{-1}((\Omega_s)_{KN_s}),
\qquad (4.38)
$$

one has

$$
G_s \subset (G_{s-1})_{KN_s} - 3\beta_s,
\qquad (4.39)
$$

where the right side is the restriction required for the momentum domain from the proof of statements (III) and (IV) below. (They evidently hold then on the smaller domain G_s.) Indeed, Equation (6.102) reads for the present domains,

$$
A_s(G_s + 3\beta_s) \subset (\Omega_{s-1})_{KN_s} - (5 + \Theta_{s-1})\beta_s.
\qquad (4.40)
$$

With $\|A_s - A_{s-1}\| < \beta_s$ from Eq. (4.20), this becomes[12]

$$
A_{s-1}(G_s + 3\beta_s) \subset (\Omega_{s-1})_{KN_s} - (4 + \Theta_{s-1})\beta_s,
\qquad (4.41)
$$

and therefore

$$
G_s + 3\beta_s \subset A_{s-1}^{-1}((\Omega_{s-1})_{KN_s}) = (G_{s-1})_{KN_s},
\qquad (4.42)
$$

which is Eq. (4.39). Let thus $b \equiv 3\beta_s$ and $\Omega_0 \equiv (\Omega_{s-1})_{KN_s}$. Then Eq. (6.96) of Lemma G5 defines $\Omega_1 \equiv \Omega_s$ as

$$
\Omega_s = (\Omega_{s-1})_{KN_s} - d_s, \qquad\qquad d_s = (5 + 7\Theta_{s-1})\beta_s.
\qquad (4.43)
$$

According to Lemma G5, A_s is a diffeomorphism of the domain $G_s = A_s^{-1}(\Omega_s)$ (see Eq. 6.98) onto Ω_s, and Eq. (4.18) holds for A_s with θ_s and Θ_s from Eq. (4.19).

---

[11] Ω_0 and Ω_1 in Lemma G5 do not correspond to the present Ω_{s-1} and Ω_s for $s = 1$. Since we use here general subscript s, this clash of notation should cause no confusion.

[12] Arnold (1963a), p. 21, has here '$G_1 + 3\beta$ is mapped into Ω_{KN},' which apparently refers to mapping by $A' = A_s$.

Replacing the variable $p_{s-1} \in G_{s-1}$ by $p_s \in G_s$, statement (I) is proved.[13] Statement (II) corresponds to Eq. (6.75) of Lemma G5 with the appropriate subscripts s added.

The proof of statements (III) and (IV) on B_s and H_s is in seven steps, (i) to (vii).

(i) *Introducing Σ_1 to Σ_4*

The canonical transformation that shifts the perturbation of the Hamiltonian from upper bound M_s to $M_{s+1} \ll M_s$ is given by

$$p = P + S_q(P,q), \qquad\qquad Q = q + S_P(P,q), \qquad (4.44)$$

with generating function $(P,q) + S(P,q)$ (the first term is the scalar product, $(P,q) = \sum_{i=1}^{n} P_i q_i$). The small number ε from the last chapter is absorbed here into S. A new line marked with '$*$' is inserted between the last two lines of Eq. (4.1),

$$
\begin{aligned}
H_{s-1}^+(p) &+ H_{s-1}^-(p,q) \\
&= H_{s-1}^+(p) + \overline{H}(p) + \tilde{H}(p,q) \\
&= H_s^+(p) + \tilde{H}(p,q) \\
&\overset{*}{=} H_s^+(P) + \Sigma_1 + \Sigma_2 + \Sigma_3 + \Sigma_4 \\
&= H_s^+(P) + H_s^-(P,Q),
\end{aligned}
\qquad (4.45)
$$

where, using $S_q = p - P$,

$$\Sigma_1 = \left(\frac{\partial H_s^+(P)}{\partial P}, S_q(P,q) \right) + \left[\tilde{H}(P,q) \right]_{N_s}, \qquad (4.46)$$

$$\Sigma_2 = H_s^+(p) - H_s^+(P) - \left(p - P, \frac{\partial H_s^+(P)}{\partial P} \right), \qquad (4.47)$$

$$\Sigma_3 = \tilde{H}(P,q) - \left[\tilde{H}(P,q) \right]_{N_s}, \qquad (4.48)$$

$$\Sigma_4 = \tilde{H}(p,q) - \tilde{H}(P,q). \qquad (4.49)$$

The goal is to show for $H_s^-(P,Q) = \Sigma_1 + \Sigma_2 + \Sigma_3 + \Sigma_4$ that

$$\left| H_s^-(P,Q) \right| = \left| \Sigma_1 + \Sigma_2 + \Sigma_3 + \Sigma_4 \right| < M_{s+1} = M_s^2 \delta_s^{-2\nu} \beta_s^{-2}. \qquad (4.50)$$

In Σ_1 to Σ_4 it is understood that p and q are replaced by P and Q using $p = p(P,Q)$ and $q = q(P,Q)$.[14] In the transition from $H_s^+(p)$ to $H_s^+(P)$ in Eq. (4.45), however, the

---

[13] Arnold (1963b), p. 152, item 2. has a combination ρ, θ', Θ', or $\rho_{s-1}, \theta_s, \Theta_s$ in our notation. This mixture of 'old' ρ with 'new' θ and Θ does not appear anywhere else in Arnold (1963a,b). It is a consequence of the somewhat artificial suppression of the perturbation of the frequency map, $A_{s-1} \to A_s$, in Arnold's Fundamental lemma on the transformation $H_{s-1}^- \to H_s^-$ of the perturbed Hamiltonian (Arnold 1963a, p. 18; Arnold 1963b, p. 146). This is avoided here by merging his Fundamental lemma and Inductive lemma into a single theorem.

[14] '[T]he variables p,q ... being replaced, after all the differentiations, by their expressions in terms of P,Q.' (Arnold 1963b, p. 149)

variable is simply renamed from p to P. The corresponding motions on phase-space tori have frequencies

$$\omega(p) = A_s(p) = \frac{\partial H_s^+(p)}{\partial p} \neq \frac{\partial H_s^+(P)}{\partial P} = A_s(P) = \omega(P), \qquad (4.51)$$

since $p \neq P$.[15] Thus

$$\Sigma_1 = \big(\omega(P), S_q(P,q)\big) + \big[\tilde{H}(P,q)\big]_{N_s}. \qquad (4.52)$$

In Eq. (4.50), $\delta_s^{-2\nu} \beta_s^{-2}$ is the inverse of very small numbers and thus counteracts the shift of perturbations from M_s to M_s^2. With M_s a sufficiently large power of δ_s, one can still achieve $M_{s+1} = M_s^{3/2}$. Then an infinite sequence of infinitesimal canonical transformations gives an integrable Hamiltonian. In the subsequent steps (ii, iv, v, vi), estimates are given for Σ_1 to Σ_4, from which Eq. (4.50) is obtained in (vii). The existence of the canonical diffeomorphism $B_s(P,Q) = (p,q)$ is shown in (iii).

In preparation for steps (ii, iv, v, vi), note that Σ_1 becomes zero for an appropriate choice of $S(P,q)$. Σ_3 is of order M_s^2 for N_s being sufficiently large. Σ_2 is the term of order $|p - P|^2 \sim M_s^2$ in a Taylor series expansion of H_s^+. Similarly, Σ_4 is of order $|p - P|\tilde{H} \sim M_s^2$. The δ_s and β_s in Eq. (4.50) are derived in (iv,v,vi).

(ii) $\Sigma_1 = 0$

The perturbation $\tilde{H}$ is expanded in a Fourier series with respect to the 2π-periodic angle variables q_1 to q_n; this Fourier series is truncated at $N_s \gg 1$. For the generating function S we choose a trigonometric polynomial of order $N_s - 1$,[16]

$$[\tilde{H}(P,q)]_{N_s} = \sum_{0<|k|<N_s} \tilde{h}_k(P)\, e^{i(k,q)}, \qquad (4.53)$$

$$S(P,q) = \sum_{0<|k|<N_s} s_k(P)\, e^{i(k,q)}. \qquad (4.54)$$

There is no term $k = 0$ (see Section 3.1): the zeroth Fourier component of $\tilde{H}$ is $\bar{H}$, and is absorbed into H_s^+ (see Eq. 4.1); and for S, only the derivative with respect to q enters σ_1, thus the zeroth Fourier component (which has no factor $e^{i(k,q)}$) drops out,[17]

$$S_q(P,q) = \sum_{0<|k|<N_s} iks_k(P)\, e^{i(k,q)} \qquad (4.55)$$

---

[15]This is not yet the shift $A_{s-1} \to A_s$ of the frequency map. The reason for the appearance of mixed variables, i.e., new P and old q in $\tilde{H}(P,q)$, is that $S = S(P,q)$: Σ_1 in Eq. (4.46) below is put to zero, and depends linearly on $\tilde{H}$ and S. '[A]s an intermediate step, we are considering H as a function of mixed variables y' [here P] and x [here q] (and this causes no problem)' (Chierchia 2009, p. 5069).

[16]We follow Arnold (1963a) and avoid writing $[S]_{N_s}$.

[17]Note that q in S_q is a derivative, whereas k in s_k is a multi-index.

with $k = (k_1, \ldots, k_n)$. Inserting Eqs. (4.53) and (4.55) in Σ_1 from Eq. (4.46) gives, because of orthogonality of the $e^{i(k,q)}$,

$$(\omega(P), ik\, s_k(P)) + \tilde{h}_k(P) = 0 \tag{4.56}$$

or

$$\boxed{\, s_k(P) = \frac{i\tilde{h}_k(P)}{(k, \omega(P))} \,} \tag{4.57}$$

In this equation, one encounters the 'small divisor' problem:

> For certain 'resonance' values ω the denominator (ω, k) is arbitrarily small for suitable k. These small denominators cast suspicion on the validity of our formal transformations. (Arnold 1963a, p. 12)

A Diophantine condition is used to exclude the resonances. Let $(\Omega_{s-1})_{KN_s}$ be defined as in Eq. (4.15), thus for $\omega \in \Omega_{s-1}$,

$$|(k, \omega)| \geq \frac{K}{|k|^{n+1}} \qquad \text{for} \qquad 0 < |k| < N_s \tag{4.58}$$

with N_s from Eq. (4.16). The corresponding domain for P is defined by the inverse frequency map A_s^{-1}. However, A_s is a diffeomorphism from G_s onto Ω_s, with $\Omega_s \subset (\Omega_{s-1})_{KN_s}$, thus A_s^{-1} is not unique on $(\Omega_{s-1})_{KN_s}$. All domain reductions applied in the following to obtain $\|H_s^-\| \leq M_{s+1}$ will leave a set that is *larger* than Ω_s. Hence the restriction of functions A_s, B_s, H_s to the domain G_s (and for A_s, to the range Ω_s) is the most stringent, and will be applied in all statements (I) to (IV) of the Fundamental theorem. To keep track of the domain reductions required to obtain $\|H_s^-\| \leq M_{s+1}$, we write

$$(G_{s-1})_{KN_s} = A_s^{-1}((\Omega_{s-1})_{KN_s}), \tag{4.59}$$

where it is understood that the momentum and frequency domains are reduced to subsets of G_s and Ω_s for which $A_s : G_s \to \Omega_s$ is a diffeomorphism.

From $H_{s-1}^-(p,q) = \bar{H}(p) + \tilde{H}(p,q)$ with $\|H_{s-1}^-\| \leq M_s$ (Eq. 4.5) and thus $\|\bar{H}\| \leq M_s$ (Eq. 4.31) one has[18] $\|\tilde{H}\| \leq 2M_s$ (triangle inequality) for $|\operatorname{Im} q| \leq \rho_{s-1}$. Equation (5.18) from Lemma A2 applies then to the Fourier coefficients of $\tilde{H}$,

$$|\tilde{h}_k| \leq 2M_s e^{-|k|\rho_{s-1}}. \tag{4.60}$$

Using Eq. (5.1) from Lemma A1 gives, with m replaced by $|k|$ and abbreviating $\nu_1 = n+1$,

$$|k|^{\nu_1} \leq \frac{\nu_1^{\nu_1}}{e^{\nu_1}} \frac{e^{|k|\delta_s}}{\delta_s^{\nu_1}}, \tag{4.61}$$

---

[18]Arnold (1963a) proves his Fundamental lemma for $\|\tilde{H}\| \leq M_s$, thus the following formulas deviate by factors of 2 from those in Arnold. To stay close to Arnold, we keep $2M_s$ without absorbing the factor 2 into constants.

where the control variable $\delta_s > 0$ is chosen to obtain the intended bounds. Inserting Eqs. (4.58, 4.60, 4.61) in Eq. (4.57) yields

$$|s_k(P)| = \frac{|\tilde{h}_k|}{|(k,\omega)|} \leq 2M_s e^{-|k|\rho_{s-1}} \frac{|k|^{\nu_1}}{K} \leq \frac{M_s e^{-|k|\rho_{s-1}}}{K} \frac{\nu_1^{\nu_1}}{e^{\nu_1}} \frac{e^{|k|\delta_s}}{\delta_s^{\nu_1}}, \qquad (4.62)$$

hence

$$|s_k(P)| \leq \frac{2M_s}{K} \frac{L_0}{\delta_s^{\nu_1}} e^{-|k|(\rho_{s-1}-\delta_s)}, \qquad (4.63)$$

where

$$L_0 = (\nu_1/e)^{\nu_1}. \qquad (4.64)$$

With the small divisors from Eq. (4.57) excluded by the Diophantine condition, the Fourier coefficients $s_k(P)$ are bounded, and S as a trigonometric polynomial is analytic. This is also true for $N_s \to \infty$, i.e., for the full Fourier series:

> With the [Diophantine] condition the coefficients $s_k = i\tilde{h}_k/(\omega,k)$ decrease in geometric progression almost as rapidly as the $\tilde{h}_k$ [...] Consequently, the series S converges for $|\mathrm{Im}\, q| < \rho$. (Arnold 1963b, pp. 98-99)

To find a bound for $|S(P,q)|$, which is used in (iii) below to obtain the diffeomorphism B_s, Eq. (5.19) from Lemma A2 is used (the 'inverse' of Eq. 5.18),

$$\text{if} \quad |f_k| \leq M e^{-|k|\rho} \quad \text{then} \quad |f(q)| < \frac{4^n M}{\delta^n} \quad \text{for} \quad |\mathrm{Im}\, q| \leq \rho - \delta, \qquad (4.65)$$

with $0 < \delta < \rho \leq 1$. This estimate for $|f(q)|$ holds for the infinite sum over all $|k|$, thus also for a polynomial with $|k| < N_s$ (there are no cancellations between different k in the proof of Eq. 5.19, see Eq. 5.29). For the present case with $f_k \equiv s_k$, $M \equiv 2M_s L_0/(K\delta_s^{\nu_1}) = M_s'$, $\rho \equiv \rho_{s-1} - \delta_s$, and treating P as a parameter, one has

$$\text{if} \quad |s_k(P)| \leq M_s' e^{-|k|(\rho_{s-1}-\delta_s)} \quad \text{then} \quad |S(P,q)| < \frac{4^n M_s'}{\delta_s^n} \quad \text{for} \quad |\mathrm{Im}\, q| \leq \rho_{s-1} - 2\delta_s, $$
$$(4.66)$$

where $2\delta_s < \rho_{s-1}$ must hold, which is true since $10\delta_s < \rho_{s-1}$ (Eq. 4.7).[19] Thus

$$|S(P,q)| < \frac{4^n}{\delta_s^n} \frac{2M_s L_0}{K\delta_s^{\nu_1}} = \frac{2M_s L_5}{K\delta_s^{\nu_2}} \qquad (4.67)$$

with $\nu_2 = 2n+1$ and

$$L_5 = 4^n L_0, \qquad (4.68)$$

---

[19]The restriction $|\mathrm{Im}\, q| \leq \rho_{s-1} - 2\delta_s$ in Eq. (4.66) is also obtained directly from the second and third line in the derivation (5.29), where an exponential

$$e^{-|k|(\rho_{s-1}-\delta_s)} e^{|k|(\rho_{s-1}-2\delta_s)} = e^{-|k|\delta_s}$$

is required for the subsequent steps.

for $P \in (G_{s-1})_{KN_s}$ and $|\mathrm{Im}\, q| \le \rho_{s-1} - 2\delta_s$.

(iii) *Diffeomorphism* B_s

By Eq. (4.10), $M_s < \delta_s^\nu K \beta_s^2$, thus

$$|S(P,q)| < \frac{2M_s L_5}{K\delta_s^{\nu 2}} < \frac{2\delta_s^\nu \beta_s^2 L_5}{\delta_s^{\nu 2}} = \frac{2\delta_s^{2n+3}\beta_s^2 L_5}{\delta_s^{2n+1}} \stackrel{(a)}{=} \frac{2\delta_s^2 \beta_s^2 L_2}{16n} \stackrel{(b)}{\le} \frac{\beta_s^2 \delta_s L_2}{16n} \stackrel{(c)}{<} \frac{\beta_s^2}{16n},$$
(4.69)

introducing in (a)

$$L_2 = 16nL_5,$$
(4.70)

and using $\delta_s < 1/2$ from Eq. (4.7) in (b) and $\delta_s L_2 < 1$ from Eq. (4.6) in (c).

The assumptions of Lemma G3 are then met, especially the bound $\|S\| \le \beta_s^2/(16n)$ in Eq. (6.37). According to this lemma, the generating function $S(P,q)$ defined on a domain $(P,q) \in G \times U$ establishes a canonical diffeomorphism $B_s(P,Q) = (p,q)$ on $G - 2\beta_s \times U - 3\beta_s$. With $P \in (G_{s-1})_{KN_s}$ and $|\mathrm{Im}\, q| \le \rho_{s-1} - 2\delta_s$, the domain restriction for $B_s(P,Q)$ from Lemma G3 is thus

$$P \in (G_{s-1})_{KN_s} - 2\beta_s, \qquad |\mathrm{Im}\, Q| \le \rho_{s-1} - 2\delta_s - 3\beta_s.$$
(4.71)

This is restricted further, to

$$P \in (G_{s-1})_{KN_s} - 2\beta_s, \qquad |\mathrm{Im}\, Q| \le \rho_{s-1} - 3\delta_s - 3\beta_s,$$
(4.72)

as required by Eq. (6.42) of Lemma G3 for the estimate of $|P - p|$. To simplify matters, Eq. (4.72) is used for all cases (I) to (IV) of Lemma G3.[20] Using $3\beta_s < 2\delta_s$ from Eq. (4.8), one has $\rho_{s-1} - 5\delta_s < \rho_{s-1} - 3\delta_s - 3\beta_s$, and Eq. (4.72) can be simplified by restriction to a smaller domain,

$$P \in (G_{s-1})_{KN_s} - 2\beta_s, \qquad |\mathrm{Im}\, Q| \le \rho_{s-1} - 5\delta_s.$$
(4.73)

Equations (6.40) to (6.42) of Lemma G3 for the diffeomorphism B_s become then

$$\|B_s - E\| \le \frac{2M_s L_5}{K\delta_s^{\nu 2}} \frac{1}{\beta_s} < \frac{\beta_s^2}{16n\beta_s} < \beta_s,$$
(4.74)

$$\|dB_s\| < 2\,|dX|, \qquad X = (P,Q),$$
(4.75)

$$|P - p| \le \frac{2M_s L_5}{K\delta_s^{\nu 2+1}}.$$
(4.76)

Then $5\delta_s < \gamma_s$ from Eq. (4.7) gives $\rho_{s-1} - 5\delta_s > \rho_{s-1} - \gamma_s > \rho_{s-1} - 3\gamma_s$ (the latter restriction is required by the following estimates), and thus statement (III) of the

---

[20] As is done in Arnold (1963a). However, he merely states (p. 19, item 3°) that $\rho_{s-1} - 5\delta_s < \rho_{s-1} - 2\delta_s - 3\beta_s$, which could be improved to the less stringent $\rho_{s-1} - 4\delta_s = \rho_{s-1} - 2\delta_s - 2\delta_s < \rho_{s-1} - 2\delta_s - 3\beta_s$ of Eq. (4.71). But what is needed is indeed $\rho_{s-1} - 5\delta_s < \rho_{s-1} - 3\delta_s - 3\beta_s$ of Eq. (4.72).

Fundamental theorem holds since $G_s \subset (G_{s-1})_{KN_s} - 2\beta_s$ by Eq. (4.39). Note from Eq. (4.74) that[21]

$$\max(|p - P|, |q - Q|) = |(p, q) - (P, Q)| = |(B_s - E)(P, Q)| \le \|B_s - E\| < \beta_s. \quad (4.77)$$

(iv) *Estimate for* Σ_2

Σ_2 from Eq. (4.47) can be estimated by the second-order term in a Taylor series. From Eq. (4.18),

$$|dA_s(P)| = \left| d\, \frac{\partial H_s^+(P)}{\partial P} \right| < \Theta_s |dP|, \quad (4.78)$$

or

$$\max_i \left(\left| \sum_{j=1}^{n} \frac{\partial^2 H_s^+(P)}{\partial P_i\, \partial P_j}\, dP_j \right| \right) \le \Theta_s\, \max_i(|dP_i|). \quad (4.79)$$

This must hold for all possible $(dP_1, \ldots, dP_n)$, thus a bound for the maximum size of $\frac{\partial^2 H_s^+}{\partial P_i \partial P_j}$ is obtained for $dP = (0, \ldots, 0, dP_k, 0, \ldots, 0)$, which gives

$$\max_i \left(\left| \frac{\partial^2 H_s^+(P)}{\partial P_i\, \partial P_k}\, dP_k \right| \right) \le \Theta_s |dP_k|. \quad (4.80)$$

Hence for all i and, putting successively $k = 1, \ldots, n$, also for all j,

$$\left| \frac{\partial^2 H_s^+(P)}{\partial P_i\, \partial P_j} \right| \le \Theta_s \quad (4.81)$$

for $P \in (G_{s-1})_{KN_s} - 2\beta_s$ and $|\mathrm{Im}\, Q| \le \rho_{s-1} - 5\delta_s$. Applying Eq. (5.46) of Lemma A4 gives, with $a \equiv P, b \equiv p$ and using Eq. (4.76) for $|P - p|$,

$$|\Sigma_2| \le \frac{n^2 \Theta_s}{2} |p - P|^2 \le \frac{n^2 \Theta_s}{2} \left(\frac{2M_s L_5}{K\, \delta_s^{\nu_2 + 1}} \right)^2 = \frac{\Theta_s L_3\, 4M_s^2}{K^2\, \delta_s^{2\nu_2 + 2}} < \frac{4M_s^2}{K^2\, \delta_s^{2\nu_2 + 3}}, \quad (4.82)$$

introducing

$$L_3 = \tfrac{1}{2} n^2 L_5^2 \quad (4.83)$$

and using

$$\delta_s \overset{(4.6)}{<} \frac{1}{2\Theta_{s-1} L_3} < \frac{1}{(1 + \delta_s)\Theta_{s-1} L_3} \overset{(4.19)}{=} \frac{1}{\Theta_s L_3}. \quad (4.84)$$

(v) *Estimate for* Σ_3

---

[21] The estimates for $|P - p|$ from Eqs. (4.74) and (4.76) differ by a factor $\delta_s / \beta_s = \delta_s^{-2}$, which is significant in Eq. (4.82).

From Eqs. (4.48) and (4.53),

$$\Sigma_3 = \tilde{H}(P,q) - [\tilde{H}(P,q)]_{N_s}$$

$$= \sum_{0<|k|<\infty} \tilde{h}_k(P)\, e^{i(k,q)} - \sum_{0<|k|<N_s} \tilde{h}_k(P)\, e^{i(k,q)}$$

$$= \sum_{|k|=N_s}^{\infty} \tilde{h}_k(P)\, e^{i(k,q)} = R_{N_s}\, \tilde{H}(P,q), \tag{4.85}$$

with R_{N_s} from Eq. (5.17) of Lemma A2. We check the assumptions of statement (III) of this lemma: first, $|\tilde{h}_k| \leq 2M_s e^{-|k|\rho_{s-1}}$ from Eq. (4.60). The required restriction of the Q-domain is obtained as follows. From Eq. (5.20) in Lemma A2, $|\mathrm{Im}\, q| \leq \rho_{s-1} - \delta_s - \gamma_s$; from Eq. (4.77), $|q-Q| < \beta_s$, thus Q has to be restricted to $|\mathrm{Im}\, Q| < \rho_{s-1} - \delta_s - \gamma_s - \beta_s$.[22] One has then

$$\rho_{s-1} - \delta_s - \gamma_s - \beta_s \overset{(a)}{>} \rho_{s-1} - 2\delta_s - \gamma_s \overset{(b)}{>} \rho_{s-1} - 2\gamma_s, \tag{4.86}$$

using $3\beta_s < 2\delta_s$ in (a) and $5\delta_s < \gamma_s$ in (b). Equation (4.73) requires $\rho_{s-1} - 5\delta_s$. Since $\rho_{s-1} - 2\gamma_s < \rho_{s-1} - 5\delta_s$, we use $|\mathrm{Im}\, Q| < \rho_{s-1} - 2\gamma_s$ as most stringent restriction of the Q-domain. Then Eq. (5.20) in Lemma A2 applies and gives, with[23]

$$L_4 = 2\,(2n/e)^n \tag{4.87}$$

and $\delta_s < L_4^{-1}$ from Eq. (4.6), and with $N_s\gamma_s = -\ln(2M_s)$ from Eq. (4.16),

$$|\Sigma_3| < \left(\frac{2n}{e}\right)^n \frac{2M_s}{\delta_s^{n+1}}\, e^{-N_s\gamma_s} < L_4\, \frac{2M_s^2}{\delta_s^{\nu_1}} < \frac{2M_s^2}{\delta_s^{\nu_1+1}}. \tag{4.88}$$

(vi) Estimate for Σ_4

Assume that the domain restriction from Eq. (4.73) applies,

$$P \in (G_{s-1})_{KN_s} - 2\beta_s, \qquad |\mathrm{Im}\, Q| \leq \rho_{s-1} - 5\delta_s. \tag{4.89}$$

Then $|p-P| < \beta_s$ from Eq. (4.77) holds for B_s defined on this domain, thus $p \in (G_{s-1})_{KN_s} - \beta_s$. Since $\tilde{H}(P,q)$ is analytic if Eq. (4.89) holds and $\|\tilde{H}\| \leq 2M_s$, the Cauchy estimate in Eq. (5.33) of Lemma A3 gives, for $i = 1, \ldots, n$,

$$\left|\frac{\partial \tilde{H}(P,q)}{\partial P_i}\right| \leq \frac{2M_s}{\beta_s}. \tag{4.90}$$

---

[22] Arnold (1963a) has here $|\mathrm{Im}\, Q| < \rho - \gamma - \delta$. This is probably a misprint (also in the Russian original), and Arnold (1963b) has instead (p. 150, item 6) $|\mathrm{Im}\, q| < \rho - \gamma - \delta$. The ultimate domain restriction is unaffected by this, $|\mathrm{Im}\, Q| < \rho_{s-1} - 5\delta_s - \gamma_s$ in Arnold (1963a,b) and here.

[23] Actually $L_4 = (2n/e)^n$ from Eq. (5.20) would suffice. The extra factor of 2 (i.e., reduction in $\delta_s < L_4^{-1}$) is possibly introduced by Arnold to have $L_4 > 1$ for $n = 1$, which is true then for L_1 to L_4 for all $n \geq 1$.

One easily sees that P and p and the straight line between them lie in $(G_{s-1})_{KN_s} - \beta_s$. Thus the Lagrange formula (5.44) of Lemma A4 applies, giving (with constant $c \equiv M_s/\beta_s$)

$$|\Sigma_4| \overset{(4.49)}{=} |\tilde{H}(p,q) - \tilde{H}(P,q)|$$

$$\overset{(5.44)}{\leq} \frac{2M_s}{\beta_s} n|p - P| \overset{(4.76)}{\leq} \frac{2M_s}{\beta_s} n \frac{2M_s L_5}{K \delta_s^{\nu_2+1}} < \frac{M_s^2 L_2}{K \beta_s \delta_s^{\nu_2+1}} < \frac{M_s^2}{K \beta_s \delta_s^{\nu_2+2}}, \quad (4.91)$$

using $L_2 = 16nL_5$ from Eq. (4.70) and $\delta_s < L_2^{-1}$.

(vii)   *Estimate for* $|\Sigma_1 + \Sigma_2 + \Sigma_3 + \Sigma_4|$

P is restricted to $P \in (G_{s-1})_{KN_s} - 2\beta_s$. For Q, $|\text{Im } Q| \leq \rho_{s-1} - 5\,\delta_s$ is required in the estimates of Σ_2 and Σ_4, and a more restrictive $|\text{Im } Q| \leq \rho_{s-1} - 2\gamma_s$ in the estimate of Σ_3.

With $\Sigma_1 = 0$, Σ_2 from (4.82), Σ_3 from (4.88), Σ_4 from (4.91) and

$$\nu = 2n + 3, \qquad \nu_1 = n + 1, \qquad \nu_2 = 2n + 1 \qquad (4.92)$$

one has

$$|H_s^-(P,Q)| = |\Sigma_1 + \Sigma_2 + \Sigma_3 + \Sigma_4| < \frac{4M_s^2}{K^2 \delta_s^{2\nu_2+3}} + \frac{2M_s^2}{\delta_s^{\nu_1+1}} + \frac{M_s^2}{K \beta_s \delta_s^{\nu_2+2}}$$

$$= \frac{M_s^2}{\beta_s^2 \delta_s^{2\nu}} \left(\frac{4\beta_s^2}{K^2} \delta_s + 2\beta_s^2 \delta_s^{3n+4} + \frac{\beta_s}{K} \delta_s^{2n+3} \right)$$

$$< \frac{M_s^2}{\beta_s^2 \delta_s^{2\nu}} \left(\delta_s + \delta_s^{3n+6} + \delta_s^{2n+3} \right) \qquad (4.93)$$

using $\beta_s < K/2$ and $3\beta_s < 2\delta_s$. With $\delta_s < 1/2$ one obtains the crucial Eqs. (4.25) and (4.28),

$$|H_s^-(P,Q)| < \frac{M_s^2}{\beta_s^2 \delta_s^{2\nu}}. \qquad (4.94)$$

The norm of H_s^- is therefore dominated by Σ_2, the deviation of the frequency map A_s from being linear. The bounds (4.26) and (4.27) for the first and second derivatives of H_s^- are obtained from the Cauchy estimate (5.33) of Lemma A3, giving

$$\left\| \frac{\partial H_s^-}{\partial x_s} \right\| < \frac{M_{s+1}}{\beta_s}, \qquad \left\| \frac{\partial^2 H_s^-}{\partial x_s^2} \right\| < \frac{2M_{s+1}}{\beta_s^2} \qquad (4.95)$$

for $x_s = (p_s, q_s)$ lying in the domain with another β_s-layer taken off,[24]

$$P \in (G_{s-1})_{KN_s} - 3\beta_s, \qquad |\text{Im } Q| < \rho_{s-1} - 2\gamma_s - \beta_s. \qquad (4.96)$$

---

[24] Arnold (1963a) has here still $P \in G_{KN} - 2\beta$. But $|H_s^-(P,Q)| < M_{s+1}$ holds only in this latter domain, thus to apply the Cauchy estimate, another β has to be taken off. This is then done in item 3° of section 3.2 of Arnold (1963a). Arnold (1963b), p. 152, uses $P \in G_{KN} - 3\beta$ to apply the Cauchy estimate to obtain his Eqs. (4.3.8), corresponding to the present Eq. (4.95).

With $\beta_s < \gamma_s$, this is simplified by the restriction to a smaller domain,

$$P \in (G_{s-1})_{KN_s} - 3\beta_s, \qquad\qquad |\text{Im } Q| < \rho_{s-1} - 3\gamma_s. \qquad (4.97)$$

From Eq. (4.39), $G_s \subset (G_{s-1})_{KN_s} - 3\beta_s$. Thus $F_s \ni (p_s, q_s)$ from Eq. (4.22), with $p \in G_s$ and $|\text{Im } q_s| \le \rho_{s-1} - 3\gamma_s$, is the appropriate domain for both $B_s(p_s, q_s)$ and $H_s(p_s, q_s)$.

Finally, the L_i introduced in the proof are indeed those from Eq. (4.11). $L_1 = 16$ ensures that $15\delta_s < 1$ in Eq. (4.7). From Eq. (4.64),

$$L_0 = e^{-v_1} v_1^{v_1} = e^{-(n+1)} (n+1)^{n+1}, \qquad (4.98)$$

and from Eq. (4.68),

$$L_5 = 4^n L_0 = 4^n e^{-(n+1)} (n+1)^{n+1}. \qquad (4.99)$$

Thus, from Eq. (4.70),

$$L_2 = 16 n L_5 = 4^{n+2} n e^{-(n+1)} (n+1)^{n+1}, \qquad (4.100)$$

and from Eq. (4.83),

$$L_3 = \tfrac{1}{2} n^2 L_5^2 = \tfrac{1}{2} 4^{2n} n^2 e^{-(2n+2)} (n+1)^{2n+2} = 2^{4n-1} n^2 e^{-(2n+2)} (n+1)^{2n+2}. \qquad (4.101)$$

At last, from Eq. (4.87),

$$L_4 = 2 \left(\frac{2n}{e} \right)^n = 2^{n+1} e^{-n} n^n. \qquad (4.102)$$

This finishes the proof of the Fundamental theorem. $\qquad\qquad\qquad\qquad\qquad \square$

4.2   INDUCTIVE THEOREM

The Inductive theorem addresses the cumulative action of s subsequent canonical transformations and frequency shifts, leading to a Hamiltonian $H_s(p_s, q_s) = H_s^+(p_s) + H_s^-(p_s, q_s)$. The central results are:

1. the perturbations $H_s^-(p_s, q_s)$ form a null-sequence, $\|H_s^-\| \to 0$,
2. the integrable Hamiltonians $H_s^+(p_s)$ have domains of finite measure.

The Inductive theorem defines sequences of functions, domains and numbers. Two subsequent members of these sequences are connected by a single canonical transformation and frequency shift from the Fundamental theorem. The accumulated domain reductions for the angle variable q starting from $|\text{Im } q_0| < \rho_0$ leave a domain $|\text{Im } q_\infty| < \rho_0/3$; and the frequency diffeomorphism A_∞ is limited by $\tfrac{1}{2} \theta_0 |dp_\infty| \le A_\infty \le 2\Theta_0 |dp_\infty|$ if initially $\theta_0 |dp_0| \le A_0 \le \Theta_0 |dp_0|$.

The present formulation of the theorem is slightly rearranged compared to that of Arnold (1963a,b), emphasizing these infinite sequences.

Inductive theorem   *Consider the functions* $H_0(p_0,q_0) = H_0^+(p_0) + H_0^-(p_0,q_0)$ *and* $A_0(p_0)$ *for a Hamilton system with n degrees of freedom, the compact complex domains* F_0, G_0, Ω_0 *and positive real numbers* $D, K, \kappa, \rho_0, \theta_0, \Theta_0, \delta_1$. *Assume that:*

(A) *Domain and bound of* A_0

The function A_0 *defined by*

$$A_0(p_0) = \frac{\partial H_0^+(p_0)}{\partial p_0} \tag{4.103}$$

is a diffeomorphism from G_0 *onto* Ω_0 *of type* D,[25] *with*

$$\theta_0 \, |dp_0| \le \|dA_0\| \le \Theta_0 \, |dp_0|, \tag{4.104}$$

where $0 < \theta_0 < 1 < \Theta_0 < \infty$.

(B) *Domain and bound of* H_0

The function $H_0(p_0,q_0)$ *is analytic on the domain*

$$F_0: \qquad p_0 \in G_0, \qquad |\mathrm{Im}\, q_0| \le \rho_0, \tag{4.105}$$

where $\rho_0 \le 1$ *and* $\mathrm{Re}\ q_0$ *is* 2π-*periodic in each component. On* F_0,

$$\|H_0^-\| \le M_1 \tag{4.106}$$

with M_1 *defined in* (D).

(C) *Smallness of* δ_1

The small number δ_1 *satisfies the following formula with* $s = 1$,[26]

$$\delta_s < \min\left\{ \delta^{(1)}\left(n, \frac{\theta_{s-1}}{2}, 2\Theta_{s-1}\right), \delta^{(2)}(n, \rho_{s-1}, \kappa),\right.$$
$$\left. \delta^{(3)}(n, \theta_{s-1}, \Theta_{s-1}, \kappa, D), \delta^{(4)}(\theta_{s-1}, \kappa) \right\}, \tag{4.107}$$

with $\delta^{(1)}$ *from Eq.* (4.6) *and, with the same arguments as in Eq.* (4.107),

$$\delta^{(2)} = \min\left[\left(\frac{\rho_{s-1}}{10}\right)^{4n}, \frac{1}{4^{4n}}, \kappa \right],$$

$$\delta^{(3)} = \min\left[\left(\frac{e}{32\,n^2 + 100n}\right)^{2n}, \frac{1}{6 + 14\Theta_{s-1}}, \frac{1}{4^{n+2}} \frac{\kappa}{nD} \left(\frac{\theta_{s-1}}{\Theta_{s-1}}\right)^{n} \right],$$

$$\delta^{(4)} = \frac{\kappa}{2 + \theta_{s-1}^{-1}}. \tag{4.108}$$

---

[25] See Eq. (8.1) in Lemma D1.

[26] The extra factors of 2 in $\theta_{s-1}/2$ and $2\Theta_{s-1}$ originate in Eqs. (4.144) and (4.146) and have nothing to do with the twos in $2\Theta_{s-1}$ and $\theta_{s-1}/(2n)$ in the definition of $\delta^{(1)}$ itself.

Let $K = \delta_1 \kappa$.

(D) *Infinite sequences of numbers*

For $s = 1, 2, \ldots$ *define seven sequences of numbers, with* $T = 8n + 24$,

$$\delta_{s+1} = \delta_s^{3/2}, \qquad \beta_s = \delta_s^3, \qquad \gamma_s = \delta_s^{\frac{1}{4n}}, \qquad M_s = \delta_s^T, \qquad (4.109)$$

$$\rho_s = \rho_{s-1} - 3\gamma_s, \qquad \theta_s = \theta_{s-1}(1 - \delta_s), \qquad \Theta_s = \Theta_{s-1}(1 + \delta_s). \qquad (4.110)$$

Then for $s = 1, 2, \ldots$ *there exist sequences* (A_s), (B_s), (H_s) *of functions and* (F_s), (G_s), (Ω_s) *of domains so that the following holds:*

(0) *Smallness of* $\delta_s, \beta_s, \gamma_s, M_s$

The δ_s *from Eq. (4.109) satisfy Eq. (4.107). The* $\delta_s, \beta_s, \gamma_s, M_s$ *from Eq. (4.109) satisfy condition* (C) *of the Fundamental theorem. For all s,*

$$\rho_s > \frac{\rho_0}{3}, \qquad \theta_s > \frac{\theta_0}{2}, \qquad \Theta_s < 2\Theta_0. \qquad (4.111)$$

(I) *Domain and bound of* A_s

The functions A_s *defined by*

$$A_s(p_s) = \frac{\partial H_s^+(p_s)}{\partial p_s} = \omega \qquad (4.112)$$

are diffeomorphisms from G_s *onto* Ω_s. *Here*

$$G_s = A_s^{-1}(\Omega_s), \qquad (4.113)$$

$$\Omega_s = (\Omega_{s-1})_{KN_s} - d_s, \qquad (4.114)$$

$$(\Omega_{s-1})_{KN_s} = \left\{ \omega \in \Omega_{s-1} : |(k, \omega)| \geq \frac{K}{|k|^{n+1}} \right\} \qquad (4.115)$$

with

$$N_s = \frac{1}{\gamma_s} \ln \frac{1}{2M_s}, \qquad (4.116)$$

$$d_s = (5 + 7\Theta_{s-1})\beta_s \qquad (4.117)$$

and $0 < |k| = |k_1| + \cdots + |k_n| < N_s$ *for* $k \in \mathbb{Z}^n$. *Furthermore,*

$$\theta_s |dp_s| < \|dA_s\| < \Theta_s |dp_s|, \qquad (4.118)$$

where

$$\theta_s = (1 - \delta_s)\theta_{s-1}, \qquad \Theta_s = (1 + \delta_s)\Theta_{s-1}. \qquad (4.119)$$

The change in A is

$$\|A_s - A_{s-1}\| < \beta_s \, \delta_s. \tag{4.120}$$

(II) *Measure of lost domain*

The difference between G_0 and G_s can be made arbitrarily small,

$$\mathrm{mes}\,(G_0 \backslash G_s) \leq \kappa \, \mathrm{mes}\,(G_0). \tag{4.121}$$

(III) *Domain and bound of B_s*

The functions $B_s : F_s \to F_{s-1}$, $B(p_s, q_s) = (p_{s-1}, q_{s-1})$ are diffeomorphisms on

$$F_s : \qquad p_s \in G_s, \qquad |\mathrm{Im}\, q_s| \leq \rho_s, \tag{4.122}$$

$F_s \subset F_{s-1} - \beta_s$, and the B_s obey the bounds $(x_s = (p_s, q_s))$

$$\|B_s - E\| < \beta_s, \qquad \qquad \|dB_s\| < 2\,|dx_s|. \tag{4.123}$$

(IV) *Domain and bound of H_s*

For $(p_s, q_s) \in F_s$ and $(p_0, q_0) = (B_1 \circ B_2 \circ \cdots \circ B_s)(p_s, q_s)$ one has

$$H_0(p_0, q_0) = H_s(p_s, q_s) = H_s^+(p_s) + H_s^-(p_s, q_s) \tag{4.124}$$

with [27]

$$\|H_s^-\| < M_{s+1}, \tag{4.125}$$
$$\|\partial H_s^- / \partial x_s\| < M_{s+1}/\beta_s, \tag{4.126}$$
$$\|\partial^2 H_s^- / \partial x_s^2\| < 2M_{s+1}/\beta_s^2. \tag{4.127}$$

where

$$M_{s+1} = M_s^2 \, \delta_s^{-2\nu} \, \beta_s^{-2}. \tag{4.128}$$

Proof   The proof is in four numbered steps: in step (i), statements (0), (I), (III) and (IV) are proved,[28] however, with Eq. (4.111) derived separately in step (ii); steps (iii) and (iv) prove statement (II).

(i) *Repeated applicability of Fundamental theorem*   We show that under the conditions of the Inductive theorem, the Fundamental theorem is applicable for all s.

Conditions (A) and (B) of the Inductive theorem guarantee that conditions (A) and (B) of the Fundamental theorem hold at iteration start $s = 1$, for functions H_0 and A_0 with bounds $\rho_0, \theta_0, \Theta_0, M_1$ on domains F_0, G_0, Ω_0.

---

[27] As before, we retain Arnold's $\|H_s^-\| < M_{s+1}$ instead of the usual $\|H_s^-\| \leq M_{s+1}$.

[28] This merges proof steps 1°, 4° and 5° of Arnold (1963a).

Conditions (C) and (D) of the Inductive theorem guarantee that condition (C) of the Fundamental theorem holds for all $s = 1, 2, \dots$. This is shown in three steps. First, $\delta_1 < \delta^{(1)}(n, \theta_0/2, 2\Theta_0)$ from Eq. (4.107) of the Inductive theorem implies that $\delta_1 < \delta^{(1)}(n, \theta_0, \Theta_0)$, as required by Eq. (4.6) of the Fundamental theorem. Second, we show by induction that: if δ_s obeys Eq. (4.107) for given $\rho_{s-1}, \theta_{s-1}, \Theta_{s-1}$, then $\delta_{s+1} = \delta_s^{3/2}$ obeys Eq. (4.107) with $\rho_s, \theta_s, \Theta_s$ from Eq. (4.110). The straightforward yet somewhat lengthy calculation is shifted to the end of this section. Third, we show that the iteration formulas for $\delta_s, \beta_s, \gamma_s, M_s$ in Eq. (4.109) of the Inductive theorem give Eqs. (4.7) to (4.10) of the Fundamental theorem. Indeed,

$$5\,\delta_s \overset{(a)}{<} 5\,\delta_s^{\frac{1}{2}}\,\delta_s^{\frac{1}{4n}} \overset{(b)}{<} \frac{5\,\gamma_s}{4^{2n}} < \gamma_s, \tag{4.129}$$

using $\delta_s < 1$ in (a) and $\gamma_s = \delta_s^{1/(4n)}$ together with [1][29] $\delta_s < 4^{-4n}$ from $\delta_s < \delta^{(2)}$ in (b). The latter two relations also give $\gamma_s < 1/4$, which is used below. From [2] $\delta_s < (\rho_{s-1}/10)^{4n}$ in $\delta^{(2)}$ follows $\gamma_s < \rho_{s-1}/10$, thus

$$3\,\gamma_s < \rho_{s-1}. \tag{4.130}$$

Moreover,

$$3\,\beta_s = 3\,\delta_s^3 \le \frac{3}{48}\,\delta_s < 2\,\delta_s. \tag{4.131}$$

With [3] $\delta_s < \kappa$ from $\delta^{(2)}$, one has

$$2\,\beta_s = 2\,\delta_s^3 < \delta_s^2 < \delta_s\,\kappa \le \delta_1\,\kappa = K. \tag{4.132}$$

Note that for $\delta_s \ll 1$, the definitions of β_s and γ_s imply $\beta_s \ll \delta_s \ll \gamma_s$. Finally, with $\nu = 2n + 3$ and $T = 8n + 24$,

$$M_s = \delta_s^T = \delta_s^{8n+24} = \delta_s^{2n+3}\,\delta_s^2\,\delta_s^6\,\delta_s^{6n+13} < \delta_s^\nu\,K\,\beta_s^2, \tag{4.133}$$

since $\delta_s^{6n+13} < 1$. We have to check that the crucial relation $M_{s+1} = M_s^2\,\delta_s^{-2\nu}\,\beta_s^{-2}$ (see Eqs. 4.28 and 4.128) agrees with $M_{s+1} = \delta_{s+1}^T$ from Eq. (4.109):

$$M_{s+1} \overset{(4.28)}{=} M_s^2\,\delta_s^{-2\nu}\,\beta_s^{-2} = \delta_s^{2T}\,\delta_s^{-2(2n+3)}\,\delta_s^{-6} = \delta_s^{16n+48}\,\delta_s^{-4n-6}\,\delta_s^{-6}$$
$$= \delta_s^{12n+36} = \delta_s^{3(8n+24)/2} = \delta_s^{3T/2} = \delta_{s+1}^T \overset{(4.109)}{=} M_{s+1}. \tag{4.134}$$

Thus the Fundamental theorem is applicable for $s = 1$ and gives (statements I, III, IV) new functions A_1, B_1, H_1 with bounds $\rho_1, \theta_1, \Theta_1, M_2$ on smaller domains F_1, G_1, Ω_1. These obey again assumptions (A) and (B) of the Fundamental theorem; assumption (C) of the latter for $\delta_2, \beta_2, \gamma_2, M_2$ is also met, as was just shown. The

---

[29]We mark the positions where the seven new conditions on δ introduced by $\delta^{(2)}, \delta^{(3)}, \delta^{(4)}$ are used with counters [1] to [7].

Fundamental theorem gives then new functions A_2, B_2, H_2 with bounds $\rho_2, \theta_2, \Theta_2$, M_3 on domains F_2, G_2, Ω_2, etc. The general scheme

$$p_{s-1}, q_{s-1}, A_{s-1}, H_{s-1}, F_{s-1}, G_{s-1}, \Omega_{s-1}, \rho_{s-1}, \theta_{s-1}, \Theta_{s-1}$$

$$\xrightarrow[\;]{B_s, \delta_s, \beta_s, \gamma_s, M_s} \; p_s, \quad q_s, \quad A_s, \quad H_s, \quad F_s, \quad G_s, \quad \Omega_s, \quad \rho_s, \quad \theta_s, \quad \Theta_s \quad (4.135)$$

establishes infinite sequences of canonical variables, functions, domains and numbers. This proves statements (I), (III) and (IV) of the Inductive theorem about the sequences (A_s), (B_s), (H_s). The above arguments also prove statement (0) except for Eq. (4.111), which is derived now.

(ii) *Bounds $\rho_0/3$ and $\theta_0/2$ and $2\Theta_0$*   To prove Eq. (4.111) we first show that

$$\delta_s + \delta_{s+1} + \cdots < 2\,\delta_s, \qquad (4.136)$$

$$\gamma_s + \gamma_{s+1} + \cdots < 2\,\gamma_s. \qquad (4.137)$$

Arnold (1963b) after his Eq. (5.3.2) mentions the little trick used to derive Eqs. (4.136) and (4.137). Write

$$\delta_{s+1} = \delta_s^{3/2} = \delta_s\,\delta_s^{\alpha} \qquad (4.138)$$

with $\alpha = 1/2$. Since $\delta_s \le \delta_1 < 4^{-4n} < 2^{-2} = 2^{(-1/\alpha)}$, one has

$$\delta_{s+1} = \delta_s^{\alpha}\,\delta_s \le \delta_1^{\alpha}\,\delta_s < \frac{\delta_s}{2}. \qquad (4.139)$$

Similarly,

$$\gamma_{s+1} = \delta_{s+1}^{\frac{1}{4n}} = \delta_s^{\frac{3}{2}\cdot\frac{1}{4n}} = \delta_s^{\frac{1}{4n}\cdot\frac{3}{2}} = \gamma_s^{\frac{3}{2}}. \qquad (4.140)$$

Thus also $\gamma_{s+1} = \gamma_s^{\alpha}\,\gamma_s$ with $\alpha = 1/2$. Since $\gamma_s \le \gamma_1 < 1/4 = 2^{(-1/\alpha)}$ as above, Eq. (4.139) also applies to γ_s,

$$\gamma_{s+1} < \frac{\gamma_s}{2}. \qquad (4.141)$$

From Eq. (4.139),

$$\delta_s + \delta_{s+1} + \delta_{s+2} + \cdots \quad < \quad \delta_s\left(1 + \tfrac{1}{2} + \tfrac{1}{4} + \tfrac{1}{8} + \cdots\right) = 2\,\delta_s, \qquad (4.142)$$

and similarly for γ_s. Then $\gamma_1 < \rho_0/10$ implies $3\gamma_1 < \rho_0/3$ and $3(\gamma_1 + \gamma_2 + \cdots) < 2\rho_0/3$, or

$$\rho_s > \rho_\infty = \rho_0 - 3(\gamma_1 - \gamma_2 - \cdots) > \frac{\rho_0}{3}. \qquad (4.143)$$

With $\gamma_1 + \gamma_2 + \cdots < 2\gamma_1$ and $\delta_1 + \delta_2 + \cdots < 2\delta_1$ and using $5\delta_1 < \gamma_1$ one obtains $\sum_s(\gamma_s + \delta_s) < \rho_0/2$. Arnold (1963b), p. 106, makes the somewhat cryptic statement that when the latter relation holds, 'all the approximations [of his method] are then possible.' He refers here to the fact that mes (Ω_∞) and thus G_∞ have (large) finite measure and that ρ_∞, the half-width of the strip for Im q_∞, is finite.

Regarding θ_s one obtains[30] with $\delta_1 < 1/4$ and $\delta_{s+1} < \delta_s$,

$$
\begin{aligned}
\theta_s &= \theta_{s-1}(1-\delta_s) \\
&= \theta_{s-2}(1-\delta_{s-1})(1-\delta_s) \\
&= \theta_0(1-\delta_1)(1-\delta_2)\ldots(1-\delta_s) \\
&> \theta_0\left(1-\frac{1}{4}\right)\left(1-\frac{1}{8}\right)\ldots\left(1-\frac{1}{2^s}\right) \\
&> 2\theta_0\prod_{n=1}^{\infty}\left(1-\frac{1}{2^n}\right) \\
&= 2\theta_0 \times 0.2887880950866024212788997219292230\ldots \\
&> \frac{\theta_0}{2},
\end{aligned}
\tag{4.144}
$$

where

$$
\phi(q) = \prod_{n=1}^{\infty}(1-q^n),
\tag{4.145}
$$

here with $q = 1/2$, is called Euler's function or Euler's product. For Θ_s,

$$
\begin{aligned}
\Theta_s &= \Theta_{s-1}(1+\delta_s) \\
&= \Theta_0(1+\delta_1)(1+\delta_2)\ldots(1+\delta_s) \\
&< \Theta_0\left(1+\frac{1}{4}\right)\left(1+\frac{1}{8}\right)\left(1+\frac{1}{16}\right)\left(1+\frac{1}{32}\right)\ldots \\
&= \Theta_0 \times 1.5894873526875811494332661924522264\ldots \\
&< 2\Theta_0.
\end{aligned}
\tag{4.146}
$$

(iii) *Estimate needed in* $\mathrm{mes}\,(G \backslash G_\infty)$ The following estimates are in preparation for step (iv). Equation (5.2) from Lemma A1 is, for $v = 4n$,

$$
\ln\frac{1}{\delta_s} \le \frac{4n}{e}\left(\frac{1}{\delta_s}\right)^{\frac{1}{4n}}.
\tag{4.147}
$$

Using that [4]

$$
\delta_s < \delta^{(3)} \le \left(\frac{e}{32n^2+100n}\right)^{2n},
\tag{4.148}
$$

---

[30]Using WolframAlpha at https://www.wolframalpha.com for the numerical computation.

one obtains

$$
\delta_s N_s^n \overset{(4.16)}{\leq} \delta_s \left(\frac{1}{\gamma_s} \ln \frac{1}{2M_s} \right)^n
$$

$$
= \delta_s \left(\delta_s^{-\frac{1}{4n}} \ln \frac{1}{2\,\delta_s^T} \right)^n
$$

$$
\overset{(\delta_s<1)}{<} \delta_s^{3/4} \left(\ln \frac{1}{\delta_s^{T+1}} \right)^n
$$

$$
= \delta_s^{3/4} (T+1)^n \left(\ln \frac{1}{\delta_s} \right)^n
$$

$$
\overset{(4.147)}{\leq} \delta_s^{3/4} \frac{(4n(T+1))^n}{e^n\,\delta_s^{1/4}}
$$

$$
\overset{(T=8n+24)}{=} \delta_s^{1/2} \left(\frac{32n^2 + 100n}{e} \right)^n
$$

$$
\overset{(4.148)}{<} 1. \tag{4.149}
$$

We introduce σ_s from Eq. (8.32),

$$
\sigma_s = \sum_{m=N_{s-1}}^{N_s-1} \frac{1}{m^2}, \tag{4.150}
$$

where $\sum_{s=1}^{\infty} \sigma_s < 2$. Then with[31] [5], [6] $\delta_s < \delta^{(3)}$, i.e.,

$$
\delta_s < \frac{1}{6 + 14\Theta_{s-1}} \quad \text{and} \quad \delta_s < \frac{1}{4^{n+2}} \frac{\kappa}{nD} \left(\frac{\theta_{s-1}}{\Theta_{s-1}} \right)^n, \tag{4.151}
$$

---

[31]Note that $\delta^{(2)}$ was used in part (ii) of the proof and $\delta^{(3)}$ in part (iii). $\delta^{(4)}$ is only used in the proof of the KAM theorem.

one has that[32]

$$\sum_{s=1}^{\infty}\left[K\sigma_s + (6+7\Theta_{s-1})\beta_s N_s^n\right]$$

$$\overset{(\beta_s=\delta_s^3)}{=}\sum_{s=1}^{\infty}\left[K\sigma_s + (6+7\Theta_{s-1})\delta_s^2\,\delta_s\,N_s^n\right]$$

$$\overset{(4.149)}{<}\sum_{s=1}^{\infty}\left[K\sigma_s + (6+7\Theta_{s-1})\delta_s^2\right]$$

$$\overset{(4.151)}{<}\sum_{s=1}^{\infty}\left[K\sigma_s + \frac{6+7\Theta_{s-1}}{6+14\Theta_{s-1}}\,\delta_s\right]$$

$$< \kappa\delta_1\sum_{s=1}^{\infty}\sigma_s + \sum_{s=1}^{\infty}\delta_s$$

$$\overset{(\kappa<1)}{<}2\,\delta_1 + 2\,\delta_1$$

$$\overset{(4.151)}{<}\frac{\kappa}{n\,4^{n+1}\,D}\left(\frac{\theta_0}{\Theta_0}\right)^n$$

$$= \frac{\kappa}{n\,2^{n+2}\,D}\left(\frac{\theta_0}{2\,\Theta_0}\right)^n$$

$$= \frac{\kappa}{\overline{D}}, \tag{4.152}$$

with (using $L = n2^{n+2}$ from Eq. 8.32)

$$\overline{D} = DL\left(\frac{2\Theta_0}{\theta_0}\right)^n. \tag{4.153}$$

(iv) *Measure of* $G_0\backslash G_s$ We show that the domain G_∞ for p_∞ has finite measure, which converges to G_0 for vanishing perturbations. Equation (4.21) from the Fundamental theorem, together with $\theta_s > \theta_0/2$, gives

$$\text{mes}\,(G_{s-1}\backslash G_s) \le \frac{1}{\theta_{s-1}^n}\,\text{mes}\,(\Omega_{s-1}\backslash\overline{\Omega}_s) < \frac{2^n}{\theta_0^n}\,\text{mes}\,(\Omega_{s-1}\backslash\overline{\Omega}_s), \tag{4.154}$$

where

$$\overline{\Omega}_s = (\Omega_{s-1})_{KN_s} - \overline{d}_s, \qquad N_s = \frac{1}{\gamma_s}\ln\frac{1}{2M_s}, \qquad \overline{d}_s = (6+7\Theta_{s-1})\beta_s. \tag{4.155}$$

Ω_{s-1} is last element of the sequence $\Omega_0,\ldots,\Omega_m,\ldots,\Omega_{s-1}$, where

$$\Omega_m = (\Omega_{m-1})_{KN_m} - d_m, \qquad N_m = \frac{1}{\gamma_m}\ln\frac{1}{2M_m}, \qquad d_m = (5+7\Theta_{m-1})\beta_m. \tag{4.156}$$

---

[32] According to Arnold (1963b), p. 106: 'We choose α [in $\gamma_s = \delta_s^\alpha$] sufficiently small so that $\beta_s N_s^n < \delta_s$.' Thus small α implies large γ, which implies relatively small N_s.

From Eq. (8.31) in Lemma D7, replacing Ω_s by $\overline{\Omega}_s$ and d_s by $\overline{d}_s$,

$$\text{mes}\left(\Omega_{s-1}\setminus\overline{\Omega}_s\right) \leq LD\left[K\sigma_s + (6+7\Theta_{s-1})\beta_s N_s^n\right]\text{mes}(\Omega_0). \tag{4.157}$$

For the measure of the start domains (compare to Eq. 6.106),

$$\text{mes}(\Omega_0) = \int_{\Omega_0} d\omega = \int_{G_0} dp_0 \left|\frac{dA_0}{dp_0}\right| \leq \Theta_0^n \int_{G_0} dp_0 = \Theta_0^n \,\text{mes}(G_0). \tag{4.158}$$

Clearly,

$$G_0\setminus G_s = (G_0\setminus G_1)\cup(G_1\setminus G_2)\cup(G_2\setminus G_3)\cup\ldots\cup(G_{s-1}\setminus G_s). \tag{4.159}$$

Since the sets in brackets have no points in common,

$$\text{mes}(G_0\setminus G_s) = \text{mes}(G_0\setminus G_1)+\text{mes}(G_1\setminus G_2)+\cdots+\text{mes}(G_{s-1}\setminus G_s). \tag{4.160}$$

Therefore,

$$
\begin{aligned}
\text{mes}(G_0\setminus G_s) &= \sum_{m=1}^{s}\text{mes}(G_{m-1}\setminus G_m)\\[4pt]
&\overset{(4.154)}{<} \frac{2^n}{\theta_0^n}\sum_{m=1}^{s}\text{mes}\left(\Omega_{m-1}\setminus\overline{\Omega}_m\right)\\[4pt]
&\overset{(4.157)}{<} \frac{2^n LD}{\theta_0^n}\sum_{m=1}^{s}\left[K\sigma_m + (6+7\Theta_{m-1})\beta_m N_m^n\right]\text{mes}(\Omega_0)\\[4pt]
&\overset{(4.152)}{<} \frac{2^n LD}{\theta_0^n}\frac{\kappa}{D}\,\text{mes}(\Omega_0)\\[4pt]
&\overset{(4.158)}{\leq} \frac{2^n LD}{\theta_0^n}\frac{\kappa}{D}\,\Theta_0^n\,\text{mes}(G_0)\\[4pt]
&\overset{(4.153)}{=} \kappa\,\text{mes}(G_0).
\end{aligned}
\tag{4.161}
$$

Thus, a fraction $\kappa < 1$ of G_0 is lost by the domain restrictions. The Inductive theorem is fully proved except for a few straightforward calculations that follow now. □

We have to show that with infinite sequences δ_{s+1}, ρ_s, θ_s, Θ_s as defined in assumption (D) of the Inductive theorem, δ_{s+1} obeys the smallness condition (4.107) in assumption (C) of the Inductive theorem for all $s = 2, 3, \ldots$ if it does so for $s = 1$.[33] Abbreviate Eq. (4.107) as

$$\delta_s < \delta^{(5)}(n, \kappa, D, \rho_{s-1}, \theta_{s-1}, \Theta_{s-1}). \tag{4.162}$$

Then the following lemma holds.

---

[33]Cf. Pöschel (2001), p. 24, in a related context: 'Then these inequalities hold inductively, provided they hold initially.'

Lemma *If*

$$\delta_{s+1} = \delta_s^{3/2},$$
$$\rho_s = \rho_{s-1} - 3\gamma_s,$$
$$\theta_s = \theta_{s-1}(1 - \delta_s),$$
$$\Theta_s = \Theta_{s-1}(1 + \delta_s), \tag{4.163}$$

and

$$\delta_1 < \delta^{(5)}(n, \kappa, D, \rho_0, \theta_0, \Theta_0), \tag{4.164}$$

then for $s = 2, 3, \ldots,$

$$\delta_s < \delta^{(5)}(n, \kappa, D, \rho_{s-1}, \theta_{s-1}, \Theta_{s-1}). \tag{4.165}$$

Proof By induction from s to $s+1$; thus only the six terms in $\delta^{(5)}$ that depend on $\rho_s, \theta_s, \Theta_s$ need to be considered. Equation (4.139) is used throughout,

$$\delta_{s+1} < \frac{\delta_s}{2}. \tag{4.166}$$

(i) $\underline{\delta < \theta/(4n)}$

For the first term in $\delta^{(1)}(n, \theta, \Theta)$ that depends on θ, the induction assumption (IA) $\delta_s < \theta_{s-1}/(4n)$ gives, together with $\delta_s < 1/2$,

$$\delta_{s+1} < \frac{\delta_s}{2} \stackrel{\text{IA}}{<} \frac{1}{2}\frac{\theta_{s-1}}{4n} < \frac{\theta_{s-1}(1 - \delta_s)}{4n} = \frac{\theta_s}{4n}. \tag{4.167}$$

(ii) $\underline{\delta < 1/(4\Theta L_3)}$

The induction assumption $\delta_s < 1/(4\Theta_{s-1}L_3)$ gives

$$\delta_{s+1} < \frac{\delta_s}{2} \stackrel{\text{IA}}{<} \frac{1}{2}\frac{1}{4\Theta_{s-1}L_3} < \frac{1}{4\Theta_{s-1}(1 + \delta_s)L_3} = \frac{1}{4\Theta_s L_3}. \tag{4.168}$$

(iii) $\underline{\delta < (\rho/10)^{4n}}$

For $\delta^{(2)}(n, \rho, \kappa)$ we perform a reverse calculation, starting with

$$\delta_s < \frac{1}{4^{4n}} < \frac{1}{(2.04\ldots)^{4n}} = \frac{1}{((10/7)^2)^{4n}} = \left(\frac{7}{10}\right)^{8n}. \tag{4.169}$$

The first inequality is valid for all s since it holds for $s = 1$. Thus

$$\delta_s^{\frac{1}{2}} < \left(\frac{7}{10}\right)^{4n} \quad \rightarrow \quad \delta_s^{\frac{3}{2}} < \left(\frac{7}{10}\right)^{4n}\delta_s \quad \rightarrow \quad \delta_s^{\frac{3}{8n}} < \frac{7}{10}\delta_s^{\frac{1}{4n}}$$

$$\rightarrow \quad \delta_s^{\frac{3}{8n}} + \frac{3}{10}\delta_s^{\frac{1}{4n}} < \delta_s^{\frac{1}{4n}} \stackrel{\text{IA}}{<} \frac{\rho_{s-1}}{10} \quad \rightarrow \quad \delta_s^{\frac{3}{8n}} < \frac{\rho_{s-1}}{10} - \frac{3}{10}\delta_s^{\frac{1}{4n}}, \tag{4.170}$$

and thus

$$\delta_{s+1} = \delta_s^{\frac{3}{2}} < \left(\frac{\rho_{s-1} - 3\delta_s^{\frac{1}{4n}}}{10}\right)^{4n} = \left(\frac{\rho_{s-1} - 3\gamma_s}{10}\right)^{4n} = \left(\frac{\rho_s}{10}\right)^{4n}. \tag{4.171}$$

(iv) $\delta < 1/(6 + 14\Theta)$

For $\delta^{(3)}(n, \theta, \Theta, \kappa, D)$ the auxiliary calculation

$$2 > 1 + \delta_s \quad \rightarrow \quad 28\Theta_{s-1} > 14\Theta_{s-1}(1 + \delta_s)$$
$$\rightarrow \quad 12 + 28\Theta_{s-1} > 6 + 14\Theta_{s-1}(1 + \delta_s) \tag{4.172}$$

gives

$$\delta_{s+1} < \frac{\delta_s}{2} \overset{\text{IA}}{<} \frac{1}{2} \frac{1}{6 + 14\Theta_{s-1}} \overset{(4.172)}{<} \frac{1}{6 + 14\Theta_{s-1}(1 + \delta_s)} = \frac{1}{6 + 14\Theta_s}. \tag{4.173}$$

(v) $\delta < \kappa/(4^{n+2}nD)(\theta/\Theta)^n$

Let

$$C = \frac{\kappa}{4^{n+2}nD}. \tag{4.174}$$

For all $n \geq 1$,

$$\left(\frac{4^{4n} - 1}{4^{4n} + 1}\right)^n > \frac{1}{2}. \tag{4.175}$$

Indeed, $(4^4 - 1)/(4^4 + 1) > 1/2$ and

$$\frac{d}{dn}\left(\frac{4^{4n} - 1}{4^{4n} + 1}\right)^n = k\left[(4^{4n} + 1)4n\,4^{4n-1} - (4^{4n} - 1)4n\,4^{4n-1}\right] > 0, \tag{4.176}$$

where k is a positive number. Thus the left side of Eq. (4.175) grows monotonically to its limit 1. Using $\delta_s < 4^{-4n}$ from $\delta^{(2)}$ one obtains

$$\delta_{s+1} < \frac{\delta_s}{2} \overset{\text{IA}}{<} \frac{C}{2}\left(\frac{\theta_{s-1}}{\Theta_{s-1}}\right)^n < C\left(\frac{1 - 4^{-4n}}{1 + 4^{-4n}}\right)^n \left(\frac{\theta_{s-1}}{\Theta_{s-1}}\right)^n$$
$$< C\left(\frac{1 - \delta_s}{1 + \delta_s}\right)^n \left(\frac{\theta_{s-1}}{\Theta_{s-1}}\right)^n = C\left(\frac{\theta_s}{\Theta_s}\right)^n. \tag{4.177}$$

(vi) $\delta < \kappa/(2 + \theta^{-1})$

For $\delta^{(4)}(\kappa, \theta)$ we start with an auxiliary calculation,

$$\theta_{s-1} < \theta_{s-1}(1 + \delta_s) \quad \rightarrow \quad \frac{2}{\theta_{s-1}} > \frac{1}{\theta_{s-1}(1 + \delta_s)} \quad \rightarrow \quad 4 + \frac{2}{\theta_{s-1}} > 2 + \frac{1}{\theta_{s-1}(1 + \delta_s)},$$

which gives

$$\frac{1}{4+\dfrac{2}{\theta_{s-1}}} < \frac{1}{2+\dfrac{1}{\theta_{s-1}(1+\delta_s)}}. \tag{4.178}$$

Therefore,

$$\delta_{s+1} < \frac{\delta_s}{2} \overset{\mathrm{IA}}{<} \frac{1}{2} \frac{\kappa}{2+\dfrac{1}{\theta_{s-1}}} \overset{(4.178)}{<} \frac{\kappa}{2+\dfrac{1}{\theta_{s-1}(1+\delta_s)}} = \frac{\kappa}{2+\dfrac{1}{\theta_s}}, \tag{4.179}$$

and the lemma is proved. □

4.3 KAM THEOREM

Having constructed domains F_s for canonical variables (p_s, q_s) and Hamiltonians $H_s^+(p_s) + H_s^-(p_s, q_s)$ with $\|H_s^-\| \leq M_{s+1} \to 0$ for $s \to \infty$ in the Inductive theorem, the KAM theorem asserts that for slightly perturbed integrable Hamiltonians, $M_1 \ll 1$, quasi-periodic phase-space trajectories $(p_0(t), q_0(t))$ exist on invariant tori, the latter forming a nowhere dense set of finite measure. These trajectories are put into one-to-one correspondence with the motion $(p_\infty, q_\infty = \omega t)$, with constants p_∞ and $\omega = \partial H_\infty^+/\partial p_\infty$, of the integrable Hamiltonian $H_\infty^+(p_\infty)$.

KAM theorem   *Let $H_0(p_0, q_0) = H_0^+(p_0) + H_0^-(p_0, q_0)$ be the Hamiltonian of a system with n degrees of freedom and trajectories $(p_0(t), q_0(t))$ given by*

$$\dot{p}_0 = -\frac{\partial H_0}{\partial q_0}, \qquad \dot{q}_0 = \frac{\partial H_0}{\partial p_0}. \tag{4.180}$$

H_0 shall be analytic on the bounded domain

$$F: \qquad p_0 \in G, \qquad |\mathrm{Im}\, q_0| \leq \rho_0 \leq 1 \tag{4.181}$$

and 2π-periodic in each component of $\mathrm{Re}\, q_0$. On F, let

$$\det \left| \frac{\partial^2 H_0^+}{\partial p_{0,i}\, \partial p_{0,j}} \right| = \det \left| \frac{\partial A_{0,i}}{\partial p_{0,j}} \right| \neq 0. \tag{4.182}$$

Then the following statement holds (with $(p_0, q_0) \in F$ in $\|H_0^-\|$):

$$\forall 0 < \kappa < 1 \quad \exists M_1(\kappa, \rho_0, G, H_0^+) > 0: \quad \|H_0^-\| \leq M_1 \to (\mathrm{I}, \mathrm{II}, \mathrm{III}, \mathrm{IV}). \tag{4.183}$$

Here:

(I) *Large invariant domain*

The phase space $\mathrm{Re}\, F$ can be decomposed according to [34]

$$\mathrm{Re}\, F = F^+ \cup F^-, \qquad F^+ \cap F^- = \emptyset, \tag{4.184}$$

---

[34]The superscripts in F^+ and F^- are only vaguely related to those in H^+ and H^-.

where F^+ is invariant and F^- is small, mes $(F^-) \leq \kappa$ mes (F). *Here F^+ is invariant if with any point (p_0, q_0), it contains the full trajectory $(p_0(t), q_0(t))$ passing through (p_0, q_0).*

(II) *Deformed tori*

F^+ *consists of invariant n-dimensional analytic tori* [35] T_ω. *The points on T_ω are given parametrically* [36] *by (see Fig. 4.1)*

$$p_0(Q) = p_\omega + f_\omega(Q),$$
$$q_0(Q) = Q + g_\omega(Q), \qquad (4.185)$$

with analytic functions f_ω, g_ω of period 2π in the angle variables $Q_1, \ldots, Q_n$. For given $\omega = A_\infty(p_\infty)$ with arbitrary $(p_\infty, 0) \in F^+$, the unique [37] *'reference' torus p_ω of the integrable Hamiltonian H_0^+ is defined by, with $A_0 = \partial H_0^+ / \partial p_0$,*

$$\boxed{p_\omega = A_0^{-1}(\omega)} \qquad (4.186)$$

(III) *Smallness of torus deformation*

$$\|f_\omega\| < \kappa, \qquad \|g_\omega\| < \kappa. \qquad (4.187)$$

(IV) *Motion on deformed torus*

The motion $(p_0(t), q_0(t))$ on the deformed torus T_ω of Eq. (4.185) as solution of Eq. (4.180) is quasi-periodic, see Eq. (1.70), with $Q_i(t) = \omega_i t$ for n incommensurable frequencies ω_i,

$$p_0 = p_0(\omega_1 t, \ldots, \omega_n t), \qquad q_0 = q_0(\omega_1 t, \ldots, \omega_n t). \qquad (4.188)$$

Remark 1 The non-degeneracy condition (4.182) is central to the KAM theorem. It ensures the existence of a diffeomorphism A from momentum space (with action variable p) to frequency space. The momentum of a pendulum scales linearly with the amplitude, which explains the term 'amplitude' in the following quotation:

---

[35] The frequencies $\omega = (\omega_1, \ldots, \omega_n)$ of motion on a torus serve to identify (or name) the torus. This is possible because the frequency map is a diffeomorphism.

[36] If the torus deformation is small enough that each ray from the coordinate origin inside the torus meets the torus in exactly one point, then a parameterization $p_0 = p_\omega + f_\omega(Q)$, $q_0 = Q$ is possible. The parameterization in Eq. (4.185) includes the case of 'wrinkled' tori (with radial overlap), see Fig. 4.1. If the deformed torus is parameterized with $q = Q$ and becomes tangent to a radial ray from the origin in some point a, then in a the speed on the torus becomes infinite, $v = dr/dt = (d\varphi/dt)(dr/d\varphi) = (d\varphi/dt)/(d\varphi/dr) = \omega/0$ in polar coordinates.

[37] The parameterization $(p_0(Q), q_0(Q))$ in Eq. (4.185) allows to choose p_ω from an infinite set of values. Thus a unique p_ω has to be selected.

This 'frequency-amplitude modulation' [more generally, $\det(H_{pp}) \neq 0$] is a genuinely nonlinear phenomenon. By contrast, in a linear system the frequencies are the same all over the phase space. As we will see, this is essential for the stability results of the KAM theory. As it is said, 'the nonlinearities have a stabilizing effect'. (Pöschel 2001, p. 4)

Remark 2 Conclusion (I) of the theorem is set-theoretic, (II) is topological (F^+ consists of tori), (III) is metrical (torus deformation is small), (IV) is dynamical (quasi-periodic motion).

Remark 3 Figure 4.1 shows the correspondence between points on the unperturbed and on the perturbed torus. Note that even for very small deformations, overlap in radial direction may occur on the deformed torus.

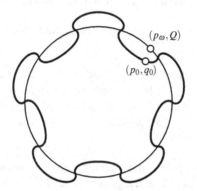

Figure 4.1 Corresponding points (p_0, q_0) of deformed and (p_ω, Q) of unperturbed torus

Remark 4 The following two sequences show the connection between the three momenta p_∞, p_ω, p_0.

$$(p_\infty, 0) \to \left\{ \begin{array}{l} \omega = A_\infty(p_\infty) \to (p_\infty, \omega t) \to \quad S_\infty(p_\infty, \omega t) \\ p_\omega = A_0^{-1}(\omega) \to (p_\omega, \omega t) \to (p_\omega + f_\omega, \omega t + g_\omega) \end{array} \right\} = (p_0(t), q_0(t))$$

Choose an arbitrary value $(p_\infty, 0) \in F^+$. With the relations

$$A_\infty(p_\infty) = \omega = A_0(p_\omega) \tag{4.189}$$

one obtains two trajectories with the same frequency ω on two different tori, for the two integrable Hamiltonians H_∞^+ and H_0^+,

$$(p_\infty, \omega t) \qquad \text{and} \qquad (p_\omega, \omega t). \tag{4.190}$$

Via the canonical transformation S_∞, $(p_0(t), q_0(t)) = S_\infty(p_\infty, \omega t)$ gives the trajectory $(p_0(t), q_0(t))$ of the perturbed system with Hamiltonian $H_0^+(p_0) + H_0^-(p_0, q_0)$ in terms of the trajectory $(p_\infty, \omega t)$ of the integrable system with Hamiltonian $H_\infty^+(p_\infty)$. The estimates $\|f_\omega\| < \kappa$ and $\|g_\omega\| < \kappa$ show that the trajectory $(p_0(t), q_0(t))$ on the deformed torus lies everywhere close to that of the unperturbed integrable system (Hamiltonian H_0^+).

Remark 5 Since the full trajectory for $-\infty < t < \infty$ lies in F^+, if one of its points does, the KAM theorem makes an asymptotic statement for all times. This is fundamentally different from statements for long but finite times, as obtained from perturbation theory of finite order.

Proof of the KAM theorem

The proof is in six steps; the last three treat the motion on phase-space tori.

(i) *Diffeomorphism A_0 on G_0* We use the following version of the inverse function theorem.

Theorem *Let U be an open domain of $\mathbb{R}^n$ and $f : U \to \mathbb{R}^n$ with $(x_1, \ldots, x_n) \mapsto (f_1(x_1, \ldots, x_n), \ldots, f_n(x_1, \ldots, x_n))$ a continuously differentiable map on U. Let $\det((\partial f_i / \partial x_j)(\bar{x})) \neq 0$ (invertible Jacobian) in $\bar{x} \in U$. Then there are open domains $U' \subset U$, with $\bar{x} \in U'$, and $V \subset \mathbb{R}^n$ so that $f : U' \to V$ is one-to-one and $f^{-1} : V \to U'$ is continuously differentiable.*

The standard *proof* applies the implicit function theorem[38] to the function $F(x, y) = y - f(x) = 0$ (with $2n$ real arguments), see Forster (1979), p. 75; Dieudonné (1969), p. 273; Sternberg (1964), p. 369. For a direct proof of the inverse function theorem, see Krantz & Parks (2013), p. 45, following Fleming (1977), p. 140. For $U \subset \mathbb{C}^n$ and analytic $f : U \to \mathbb{C}^n$, the inverse $f^{-1} : V \to U'$ is also analytic.

In the present context, $U \equiv G$ and $f \equiv A_0$. Then for $F(\omega, p_0) = \omega - A_0(p_0) = 0$, non-degeneracy $\det(\partial A_{0,i}/\partial p_{0,j}) \neq 0$ implies that a smooth inverse A_0^{-1} exists on some open domain Ω, i.e., $\omega = A_0(A_0^{-1}(\omega))$.

Applying the theorem to all (or certain) points $p_0 \in G$ one obtains open sets U_{p_0} containing p_0, which form an open covering of G. Let G_0 be a *closed* subset of G; we assume that $\mathrm{mes}(G_0) = \mathrm{mes}(G)$.[39] The U_{p_0} are an open covering of the bounded and closed and thus compact set G_0; hence there is a finite number m' of the U_{p_0}, call them U_1 to $U_{m'}$, that cover G_0. We assume that U_1 to $U_{m'}$ do not overlap.

For given numbers $0 < \theta_0 < 1$ and $1 < \Theta_0 < \infty$ exclude from U_1 to $U_{m'}$ all points for which

$$\theta_0 |dp_0| \leq |dA_0| \leq \Theta_0 |dp_0| \tag{4.191}$$

---

[38] Definition of, e.g., a surface $z = f(x, y)$ in $\mathbb{R}^3$ by an equation $F(x, y, z) = 0$.

[39] Alternatively, $\mathrm{mes}(G \backslash G_0)$ shall be arbitrarily small. Note that with G_0, also F_0 is defined.

does *not* hold, i.e., where A_0 has too small or too large slope, and include these points in $G^- = G - G_0$. Since $\det(\partial A_0/\partial p_0) \neq 0$, one can achieve for any given $0 < \kappa < 1$ that $\mathrm{mes}\,(G^-) < (\kappa/2)\,\mathrm{mes}\,(G)$ by making θ_0 sufficiently small and Θ_0 sufficiently large. Excluding points with too small or too large A_0-slope from U_1 to $U_{m'}$ will give m subsets $G_{0,1}$ to $G_{0,m}$ of G_0, where $G = G_{0,1} \cup \ldots \cup G_{0,m} \cup G^-$.

Following Arnold (1963b), p. 155, 'we shall assume that the [KAM theorem] is proved for each domain $G_{0,i}$ separately.'[40] It is not possible to consider $G_{0,1} \cup \ldots \cup G_{0,m}$ at once, since this is not a connected domain, thus standard theorems from complex function theory do not apply. On each domain $G_{0,1}$ to $G_{0,m}$, the map $A_{0,i}$ is a diffeomorphism with bounded slope, as required in condition (A) of the Inductive theorem.[41]

(ii) M_1 *and* δ_1 There arises a subtle issue here, handled with mastery by Arnold (1963a,b) but not explained in great detail:

If we can find $M_1(\kappa,\rho_0,G_{0,i},H_0^+)$ in each of the domains $G_{0,i}$, then

$$M_1(\kappa,\rho_0,G,H_0^+) = \min_i M_1\left(\frac{\kappa}{2m},\rho_0,G_{0,i},H_0^+\right) \qquad (4.192)$$

gives the proof of Theorem 1 [the KAM theorem]. (Arnold 1963a, p. 23; symbols adapted)

Clearly the minimum of the M_1 on the right side has to be taken, to make the fractional loss of measure for each of the m domains $G_{0,i}$ smaller than $\kappa/(2m)$. The reason for the factor 2 in $\kappa/(2m)$ is also clear: the measure of domain lost from G (i.e., the complement of A_s-preimages of $\Omega_s = (\Omega_{s-1})_{KN_s} - d_s$) shall be smaller than $(\kappa/2)\,\mathrm{mes}\,(G)$, and together with the loss $\mathrm{mes}\,(G^-) < (\kappa/2)\,\mathrm{mes}\,(G)$ from above (regions where A_0 has too large or too small slope) gives a total loss $\mathrm{mes}\,(G^-) < \kappa\,\mathrm{mes}\,(G)$.[42]

The reason for the factor m in $\kappa/(2m)$ in Eq. (4.192) is as follows. We demand that

$$\frac{\mathrm{mes}\left(G_{0,1}^-\right) + \cdots + \mathrm{mes}\left(G_{0,m}^-\right)}{\mathrm{mes}\,(G)} \leq \frac{\kappa}{2}. \qquad (4.193)$$

Since none of the $G_{0,1}$ to $G_{0,m}$ is special over the others, one must have

$$\frac{\mathrm{mes}\left(G_{0,i}^-\right)}{\mathrm{mes}\,(G)} \leq \frac{\kappa}{2m} \qquad (4.194)$$

[40] A corresponding statement is missing in Arnold (1963a).

[41] 'We shall henceforth suppose that the map A_0 [...] is a diffeomorphism of $G_{0,i}$.' (Arnold 1963b, pp. 154-155, two sentences merged)

[42] The measure of F- and G-domains differs by a factor $(2\pi)^n$ only, and one can switch freely between the two.

for $i = 1, \ldots, m$, thus

$$\frac{\mathrm{mes}\left(G_{0,i}^{-}\right)}{\mathrm{mes}\left(G_{0,i}\right)} \frac{\mathrm{mes}\left(G_{0,i}\right)}{\mathrm{mes}\left(G\right)} \leq \frac{\kappa}{2m}. \tag{4.195}$$

The KAM theorem states that for given $0 < \kappa < 1$, a perturbation of the Hamiltonian that is smaller than $M_1(\kappa, \rho_0, G, H_0^{+})$ leads to a loss $\mathrm{mes}\left(G^{-}\right) \leq \kappa \,\mathrm{mes}\left(G\right)$ of invariant domain (filled by tori). In the statement and proof of the KAM theorem for the individual $G_{0,i} \subset G$ replacing the overall domain G, the function $M_1(\kappa, \rho_0, G_{0,i}, H_0^{+})$ introduced in Eq. (4.183) does not refer to G (does not 'know' G), but only to $G_{0,i}$. The bound M_1 derived for $G_{0,i}$ must hold even when $\mathrm{mes}\left(G_{0,j}\right) \to 0$ for all $j \neq i$. Thus, only $\mathrm{mes}\left(G_{0,i}\right)/\mathrm{mes}\left(G\right) \leq 1$ can be assumed in Eq. (4.195), giving

$$\frac{\mathrm{mes}\left(G_{0,i}^{-}\right)}{\mathrm{mes}\left(G_{0,i}\right)} \leq \frac{\kappa}{2m}, \tag{4.196}$$

as used in Eq. (4.192).

We also note here that the $A_{0,i}$ from the inverse function theorem are *local* diffeomorphisms only. The phase-space trajectory may pass along different branches of the global map A_0 (set union of the $A_{0,i}$). All these branches are determined by H_0^{+}, thus $M_1(\kappa, \rho_0, G_{0,i}, H_0^{+})$ is well-defined. There is also a *global* version of the inverse function theorem, called Hadamard's theorem. Quite astonishingly, besides a non-vanishing Jacobian in the full domain considered, $\det(\partial f/\partial x) \neq 0$, it requires for the existence of a *bijective* f only that f is proper. A function f is *proper* if $f^{-1}(V)$ is compact whenever V is compact. For the proof of Hadamard's theorem, see Krantz & Parks (2013), p. 125.

For the given $\kappa, \rho_0, \theta_0, \Theta_0$ choose δ_1 according to (cf. Eq. 4.107)

$$\delta_1 < \min\left\{\delta^{(1)}\left(n, \frac{\theta_0}{2}, 2\Theta_0\right), \delta^{(2)}\left(n, \rho_0, \frac{\kappa}{2m}\right),\right.$$
$$\left. \delta^{(3)}\left(n, \theta_0, \Theta_0, \frac{\kappa}{2m}, D\right), \delta^{(4)}\left(\theta_0, \frac{\kappa}{2m}\right)\right\}. \tag{4.197}$$

Let again $T = 8n + 24$ and put $\delta_1 = \min\left[\delta_1, M_1^{1/T}\right]$ and $M_1 = \delta_1^{T}$. Then assumptions (A), (B), (C) of the Inductive theorem hold.

(iii) *Finite phase-space domain* Together with the infinite sequences of numbers $\delta_s, \beta_s, \gamma_s, M_s, \rho_s, \theta_s, \Theta_s$ defined in statement (D) of the Inductive theorem, one obtains sequences (F_s) and (B_s) of domains and canonical diffeomorphisms, respectively. Their convergence to a domain F_∞ and a map $S_\infty = B_1 \circ B_2 \circ \cdots$ is shown in Lemma C1. This lemma uses the following assumptions, where d_s of the lemma is replaced here by β_s,[43]

---

[43] The following assumptions (A) to (D) are new and not related to (A) to (D) from the Inductive Theorem above.

(A) $|B_s - E| \leq \beta_s$,

(B) $F_s \subset F_{s-1} - \beta_s$,

(C) $\|dB_s\| \leq 2 |dx_s|$,

(D) $\beta_s \leq c4^{-s}$.

Assumptions (A), (B), (C) hold because of statement (III) of the Inductive theorem and (D) is shown as follows:

$$\beta_s = \delta_s^3 < \left(\frac{\delta_{s-1}}{2}\right)^3 < \cdots < \left(\frac{\delta_1}{2^{s-1}}\right)^3 \overset{(a)}{<} \frac{1}{4^{12} \, 2^{3s-3}} \leq \frac{1}{4^{11}} \frac{1}{4^s}, \qquad (4.198)$$

using $\delta < \delta^{(2)} \leq 4^{-4n}$ from Eq. (4.108) in (a). Thus Lemma C1 applies and shows that S_s converges uniformly to a map S_∞ on the compact set $F_\infty = \bigcap_{s \geq 1} F_s$. In more detail, $p_\infty \in G_\infty = \bigcap_{s \geq 1} G_s$ and $q_\infty \in ([0, 2\pi] \times [-i\rho_\infty, i\rho_\infty])^n$ with $\rho_\infty \geq \rho_0/3$.

From conclusion (II) of the Inductive theorem,

$$\mathrm{mes}\,(G_\infty) \geq (1 - \kappa)\,\mathrm{mes}\,(G_0). \qquad (4.199)$$

The following argument is new: the S_s are canonical transformations and thus preserve phase-space measure, see Section 1.4. Therefore,

$$\mathrm{mes}\,(S_s F_\infty) = \mathrm{mes}\,(F_\infty). \qquad (4.200)$$

According to Lemma C4, for a compact region F_∞ and a sequence (S_s) of maps $S_s : F_\infty \rightarrow S_s F_\infty$ that converge uniformly to a map $S_\infty : F_\infty \rightarrow S_\infty F_\infty$, one has

$$\mathrm{mes}\,(S_\infty F_\infty) \geq \overline{\lim}\ \mathrm{mes}\,(S_s F_\infty). \qquad (4.201)$$

Note here that $S_s : F_s \rightarrow F_0$. Thus $S_s F_\infty$ is well-defined since $F_s \subset F_{s-1}$ and $S_s F_\infty \subset S_s F_s$. One has then

$$\mathrm{mes}\,(S_\infty F_\infty) \overset{(4.201)}{\geq} \overline{\lim}\ \mathrm{mes}\,(S_s F_\infty)$$

$$\overset{(4.200)}{=} \mathrm{mes}\,(F_\infty)$$

$$= (2\pi)^n\,\mathrm{mes}\,(G_\infty)$$

$$\overset{(4.199)}{\geq} (2\pi)^n\,(1 - \kappa)\,\mathrm{mes}\,(G_0)$$

$$= (1 - \kappa)\,\mathrm{mes}\,(F_0). \qquad (4.202)$$

The (Cantor) set $S_\infty F_\infty \subset F_0$ is that part of the original phase space F_0 of the KAM theorem in which the motion is quasi-periodic on invariant tori. Put $F^+ = S_\infty F_\infty$; then F^+ has finite measure, which approaches the full measure of F for $\kappa \rightarrow 0$.

(iv) *Invariance* We use Eq. (4.120) from statement (I) of the Inductive theorem, $\|A_s - A_{s-1}\| < \beta_s \delta_s$, to show that the sequence of frequency diffeomorphisms A_s

converges to a map A_∞ on G_∞. With $\beta_s = \delta_s^3$, $\delta_{s+1} < \delta_s/2$ and $\delta_s < 4^{-4}$ one has

$$
\|A_s - A_\infty\| = \left\| \sum_{m=s+1}^{\infty} (A_m - A_{m-1}) \right\| \le \sum_{m=s+1}^{\infty} \|A_m - A_{m-1}\| < \sum_{m=s+1}^{\infty} \beta_m \delta_m =
$$

$$
= \sum_{m=s+1}^{\infty} \delta_m^4 < \delta_{s+1}^4 \left(1 + \frac{1}{2^4} + \frac{1}{2^8} + \cdots \right) =
$$

$$
= \frac{16}{15} \delta_{s+1}^4 = \frac{16}{15} \delta_{s+1} \beta_{s+1} < \frac{1}{2} \beta_{s+1} \to 0. \tag{4.203}
$$

Let Y_s by the vector fields (with $2n$ components)

$$
Y_s = \left(-\frac{\partial H_s}{\partial q_s}, \frac{\partial H_s}{\partial p_s} \right). \tag{4.204}
$$

Then

$$
\dot{x}_s = Y_s(x_s) \tag{4.205}
$$

are the equations of motion $\dot{p}_s = -\partial H_s/\partial q_s$ and $\dot{q}_s = \partial H_s/\partial p_s$. Our goal is to obtain the solution

$$
\dot{x}_0 = Y_0(x_0) \tag{4.206}
$$

of the perturbed system with Hamiltonian $H_0(p_0, q_0)$. The H_s and x_s for different s are related by canonical transformations. Thus, if x_s is the solution of the equations of motions for H_s, then

$$
x_0(t) = S_s(x_s(t)). \tag{4.207}
$$

We show that $s \to \infty$ applies in this equation,

$$
x_0(t) = \lim_{s \to \infty} S_s(x_s(t)). \tag{4.208}
$$

Since $H_\infty = H_\infty(p_\infty)$ is integrable, one has

$$
Y_\infty = (0, \omega) \qquad \text{with} \qquad \omega = A_\infty(p_\infty). \tag{4.209}
$$

The solution of $\dot{x}_\infty(t) = Y_\infty(x_\infty(t))$ is $x_\infty(t) = (p_\infty, q_\infty(t))$ with $p_\infty = \text{const}$ and $q_\infty(t) = \omega t$ (ω a constant, the initial phase is set to zero). This is a straight line[44] in the n-dimensional cube of side length 2π of torus angle coordinates. The frequencies ω_i are incommensurable, since resonances are avoided by a Diophantine condition. Thus the motion is quasi-periodic and fills the torus uniformly. We will show that for initial conditions $x_\infty(0) \in F_\infty$ and for $x_0(t)$ from Eq. (4.208),

$$
x_\infty(t) \in F_\infty \qquad \text{and} \qquad x_0(t) \in F^+ \tag{4.210}
$$

---

[44]This is mentioned in Arnold (1963b), but not in Arnold (1963a).

for all t. Note that the momentum p_ω from Eq. (4.186) is only introduced in (v) below.

We use Lemma C3 and perform the limit $s \to \infty$. Of the assumptions (A) to (G) of this lemma, (A) to (D) were already established in (iii) above. The remaining assumptions are, replacing again d_s from the lemma by β_s and the vector field X_s by Y_s:

(E) $\quad |Y_\infty(x_\infty) - Y_s(x_\infty)| \le \beta_{s+1} \quad$ for $\quad x_\infty \in F_\infty.$

(F) $\quad x_\infty(t) = x_\infty(0) + Y_\infty t \in F_\infty \quad$ for $\quad 0 \le t \le 1.$

(G) $\quad |\partial Y_s(x_s)/\partial x_s| \le \Theta \quad$ (Θ independent of s) for $x_s \in F_s.$

Proof of (E) An 'intermediary' vector field $\overline{Y}_s$ is introduced by

$$\overline{Y}_s(x_s) = \left(0, A_s(p_s)\right). \tag{4.211}$$

Equation (4.126) of the Inductive theorem is used to obtain the following (weak) bound,

$$\left\| \frac{\partial H_s^-}{\partial x_s} \right\| < \frac{M_{s+1}}{\beta_s} = \frac{\delta_{s+1}^{8n+21}}{\delta_s^3} \delta_{s+1}^3 < \frac{1}{2} \beta_{s+1}. \tag{4.212}$$

Together with $\partial H_s^+(p_s)/\partial q_s = 0$ and $A_s = \partial H_s^+(p_s)/\partial p_s$ this gives

$$
\begin{aligned}
\left\| Y_s - \overline{Y}_s \right\| &= \left\| \left(-\frac{\partial H_s}{\partial q_s}, \frac{\partial H_s}{\partial p_s} \right) - \left(0, A_s(p_s)\right) \right\| \\
&= \left\| \left(-\frac{\partial H_s^-}{\partial q_s}, A_s(p_s) + \frac{\partial H_s^-}{\partial p_s} \right) - \left(0, A_s(p_s)\right) \right\| \\
&= \left\| \left(\frac{\partial H_s^-}{\partial q_s}, \frac{\partial H_s^-}{\partial p_s} \right) \right\| = \left\| \frac{\partial H_s^-}{\partial x_s} \right\| < \frac{1}{2}\beta_{s+1},
\end{aligned} \tag{4.213}
$$

and Eq. (4.203) yields

$$\left\| Y_\infty - Y_s \right\| \le \left\| Y_\infty - \overline{Y}_s \right\| + \left\| \overline{Y}_s - Y_s \right\| = \left\| A_\infty - A_s \right\| + \left\| \overline{Y}_s - Y_s \right\| < \beta_{s+1}. \tag{4.214}$$

Proof of (F) With $x_\infty(0) = (p_\infty(0), q_\infty(0)) \in F_\infty$ and $Y_\infty = (0, \omega) = $ const one has $x_\infty(t) = (p_\infty(0), \omega t + q_\infty(0))$, where $p_\infty(0) \in G_\infty$ is constant, and $\omega t + q_\infty(0) \in [0, 2\pi]^n$, thus $x_\infty(t) \in F_\infty$ at *all* times (not just $0 \le t \le 1$).

Proof of (G) Abbreviating $H_q = \partial H/\partial q$, etc., and suppressing a subscript s during the calculation, one has

$$\left\| \frac{\partial Y_s}{\partial x_s} \right\| = \left\| \begin{pmatrix} -H_{q_i p_j} & -H_{q_i q_j} \\ H_{p_i p_j} & H_{p_i q_j} \end{pmatrix} \right\|_M \tag{4.215}$$

$$= \max_{(dp_i, dq_i)} \left(\frac{1}{\|(dp_i, dq_i)\|_V} \left\| \begin{pmatrix} -H_{q_i p_j}^- & -H_{q_i q_j}^- \\ [H^+ + H^-]_{p_i p_j} & H_{p_i q_j}^- \end{pmatrix} \cdot \begin{pmatrix} dp_j \\ dq_j \end{pmatrix} \right\|_V \right)$$

$$= \max_{(dp_i, dq_i)} \left(\frac{1}{\|(dp_i, dq_i)\|_V} \max \left\{ \left\| \sum_j (H_{q_i p_j}^- \, dp_j + H_{q_i q_j}^- \, dq_j) \right\|_V , \right.\right.$$

$$\left.\left. \left\| \sum_j ([H^+ + H^-]_{p_i p_j} \, dp_j + H_{p_i q_j}^- \, dq_j) \right\|_V \right\} \right),$$

where subscripts M and V refer, respectively, to the matrix and vector norms. Since $\|H^-\| \ll \|H^+\|$, the first component of the second 'max' can be dropped, and dropping also the letter V for the remaining vector norms, we obtain

$$\left\| \frac{\partial Y_s}{\partial x_s} \right\| = \max_{(dp_i, dq_i)} \frac{\left\| \sum H_{p_i p_j}^+ \, dp_j + \sum H_{p_i p_j}^- \, dp_j + \sum H_{p_i q_j}^- \, dq_j \right\|}{\|(dp_i, dq_i)\|}$$

$$\le \max_{(dp_i, dq_i)} \frac{\left\| \sum (A_i)_{p_j} \, dp_j \right\| + \left\| \sum H_{p_i p_j}^- \, dp_j + \sum H_{p_i q_j}^- \, dq_j \right\|}{\|(dp_i, dq_i)\|}$$

$$= \max_{(dp_i, dq_i)} \frac{\| dA_i \| + \left\| \sum H_{p_i p_j}^- \, dp_j + \sum H_{p_i q_j}^- \, dq_j \right\|}{\|(dp_i, dq_i)\|}$$

$$\overset{(a)}{\le} \max_{(dp_i, dq_i)} \frac{\overline{\Theta} \, \| dp_i \| + \left\| \delta \sum dp_j + \delta \sum dq_j \right\|}{\|(dp_i, dq_i)\|}$$

$$\le \max_{(dp_i, dq_i)} \frac{\overline{\Theta} \, \max(|dp_i|) + n \, \delta \max(|dp_i|) + n \, \delta \max(|dq_i|)}{\|(dp_i, dq_i)\|}$$

$$\le \max_{(dp_i, dq_i)} \frac{(\overline{\Theta} + n \delta + n \delta) \max(|dp_i|, |dq_i|)}{\max(|dp_i|, |dq_i|)}$$

$$= 2n\delta_s + \overline{\Theta} \; < \; n + \overline{\Theta}, \tag{4.216}$$

reinstalling a subscript s in the last line. In (a), a weak bound obtained again from Eq. (4.127) was used,

$$\left\| \frac{\partial^2 H_s^-}{\partial x_s^2} \right\| < \frac{2M_{s+1}}{\beta_s^2} = \frac{2\delta_{s+1}^{8n+24}}{\delta_s^6} < \delta_s. \tag{4.217}$$

Since $n + \overline{\Theta}$ is independent of s, (G) is fulfilled.[45] Thus Lemma C3 applies, and Eq. (7.43) gives Eq. (4.208),

---

[45]Arnold (1963b), p. 161, uses $\delta_s < \theta_{s-1}/(2n)$ from Eq. (4.6) and $\overline{\Theta} < 2\Theta_0$ to obtain $\|\partial Y_s/\partial x_s\| \le 1 + \overline{\Theta} < 3\Theta_0$.

$$\boxed{x_0(t) = \phi_0^t(S_\infty(x_\infty(0))) = S_\infty(x_\infty(t)) = S_\infty(x_\infty(0) + Y_\infty t)} \tag{4.218}$$

where $x_0(t)$ is the trajectory of the perturbed Hamiltonian in the KAM theorem, $\dot{x}_0 = Y_0(x_0)$, with initial conditions $x_0(0) = S_\infty(x_\infty(0)) \in F^+$.

According to Lemma C3, Eq. (4.218) holds for finite time intervals $0 \le t \le \tau = (1 + n + \overline{\Theta})^{-1}$, i.e., for a segment of the trajectory $x_0(t)$. The whole trajectory is obtained by joining segments, using the end point at $t = \tau$ of a segment $x_0^{(n)}$ as the start point at $t = 0$ of segment $x_0^{(n+1)}$. By uniqueness of trajectories in phase space, the $x_0^{(n)}$ for $n = 1, 2, \ldots$ form *one* smooth trajectory. It was shown above that $x_\infty(t)$ lies in F_∞ for all $t > 0$ if $x_\infty(0)$ does. With the definition $F^+ = S_\infty F_\infty$ one obtains $x_0(t) = S_\infty(x_\infty(t)) \in F^+$ for all $t \ge 0$. 'Thus the set F^+ is invariant' (Arnold 1963b, p. 161).

(v) *Smallness of torus deformation* With $A_0(p_0) = \partial H_0^+(p_0)/\partial p_0$, define p_ω as in Eq. (4.189) by

$$A_0(p_\omega) = \omega = A_\infty(p_\infty), \tag{4.219}$$

where $p_\infty \in G_\infty$. Since A_0 is a diffeomorphism, $p_\omega = A_0^{-1}(A_\infty(p_\infty))$ is well-defined. Then

$$p_\omega = \text{const}, \quad Q = \omega t \tag{4.220}$$

(assuming initial conditions $Q = 0$ at $t = 0$) is a quasi-periodic motion[46] on a phase-space torus of the unperturbed Hamiltonian H_0^+. The quasi-periodic motion on the invariant torus T_ω of the perturbed Hamiltonian $H^+(p) + H^-(p, q)$ is given by Eq. (4.185). From Eq. (4.203),

$$|A_0(p_\infty) - A_\infty(p_\infty)| < \beta_1. \tag{4.221}$$

From Eqs. (4.219) and (4.221),

$$|A_0(p_\infty) - A_0(p_\omega)| = |A_0(p_\infty) - A_\infty(p_\infty)| < \beta_1. \tag{4.222}$$

We use Lemma G4: Eq. (6.63) reads here $\theta_0 |dp_\omega| \le |dA_0(p_\omega)| \le \Theta_0 |dp_\omega|$, and the first subset relation in Eq. (6.64) becomes, with $\varepsilon = |p_\infty - p_\omega|$,

$$\overline{B}_{\theta_0 |p_\infty - p_\omega|}(A_0(p_\omega)) \subset A_0(\overline{B}_{|p_\infty - p_\omega|}(p_\omega)). \tag{4.223}$$

From Eq. (4.222),

$$A_0(p_\infty) \in \overline{B}_{\beta_1}(A_0(p_\omega)). \tag{4.224}$$

Thus from Eq. (4.223),

$$A_0(p_\infty) \in A_0(\overline{B}_{\beta_1/\theta_0}(p_\omega)), \tag{4.225}$$

---

[46]Thus the trajectory is dense on the torus and the choice $Q(0) = 0$ is no restriction.

which gives

$$p_\infty \in \bar{B}_{\beta_1/\theta_0}(p_\omega) \tag{4.226}$$

and

$$|p_\infty - p_\omega| \leq \frac{\beta_1}{\theta_0}. \tag{4.227}$$

We repeat here the calculation of $\|S_\infty - E\|$ from Eq. (7.10) of Lemma C1 in a slightly different way, using $\|S_s - S_{s-1}\| < 2^s \beta_s$ from Eq. (7.5) (the d_s in this equation becomes here β_s). With $\beta_s = \delta_s^3$ and $\delta_{s+1} < \delta_s/2$ one obtains

$$
\begin{aligned}
\|S_\infty - E\| &\leq \|B_1 - E\| + \sum_{s=2}^{\infty} \|S_s - S_{s-1}\| < \beta_1 + \sum_{s=2}^{\infty} 2^s \beta_s \\
&< \beta_1 + \frac{2^2}{2^3}\,\delta_1^3 + \frac{2^3}{2^3 2^3}\,\delta_1^3 + \frac{2^4}{2^3 2^3 2^3}\,\delta_1^3 + \cdots \\
&= \beta_1 + \frac{1}{2}\,\delta_1^3 + \frac{1}{2^3}\,\delta_1^3 + \frac{1}{2^5}\,\delta_1^3 + \cdots \\
&< \beta_1 + \frac{1}{2}\,\delta_1^3 + \frac{1}{2^2}\,\delta_1^3 + \frac{1}{2^3}\,\delta_1^3 + \cdots \\
&= \beta_1 + \delta_1^3 = 2\beta_1.
\end{aligned}
\tag{4.228}
$$

For $(p_\infty, q_\infty) \in F_\infty$ then

$$
\begin{aligned}
&\left| S_\infty(p_\infty, q_\infty) - (p_\omega, q_\infty) \right| \\
&\quad \leq \left| S_\infty(p_\infty, q_\infty) - (p_\infty, q_\infty) \right| + \left| (p_\infty, q_\infty) - (p_\omega, q_\infty) \right| \\
&\quad < 2\beta_1 + \frac{\beta_1}{\theta_0} \overset{(a)}{<} \kappa,
\end{aligned}
\tag{4.229}
$$

using $\beta_1/\delta_1 < 1$ and $(2 + 1/\theta_0) < \kappa/\delta_1$ in (a), from $\delta_1 < \delta^{(4)} = \kappa/(2 + 1/\theta_0)$ in Eq. (4.108), which is the last of the new conditions on δ_s from $\delta^{(2)}$ to $\delta^{(4)}$ of the Inductive theorem: [7].

(vi) *Torus motion* Write Eq. (4.218) as

$$(\tilde{p}_0(t), \tilde{q}_0(t)) = S_\infty(p_\infty, q_\infty(t)), \tag{4.230}$$

with constant $p_\infty \in G_\infty$. The time dependence on the right side is via $q_\infty(t)$ alone, thus one can introduce new functions

$$p_0 = \tilde{p}_0 \circ q_\infty^{-1} \qquad \text{and} \qquad q_0 = \tilde{q}_0 \circ q_\infty^{-1} \tag{4.231}$$

so that

$$p_0(q_\infty(t)) = \tilde{p}_0(t) \qquad \text{and} \qquad q_0(q_\infty(t)) = \tilde{q}_0(t). \tag{4.232}$$

Note that if $n \geq 2$ and *all* possible resonances are avoided for $s \to \infty$, i.e., if the motion is *quasi-periodic* with non-rational ω_i/ω_j for $i \neq j$, then the mapping $t \to q_\infty$

from $\mathbb{R}_+$ to the cube $[0, 2\pi]^n$ is injective and thus invertible, although the individual q_i are periodic. Equation (4.230) becomes then, with the new variable $Q = q_\infty(t)$,

$$(p_0(Q), q_0(Q)) = S_\infty(p_\infty, Q). \tag{4.233}$$

This can be written as in Eq. (4.185),

$$p_0(Q) = p_\omega + f_\omega(Q),$$
$$q_0(Q) = Q + g_\omega(Q), \tag{4.234}$$

with the definition of p_ω from above. If $\|f_\omega\|$ and $\|g_\omega\|$ are small, this describes motion on a deformed torus T_ω. Indeed, for arbitrary Q and using Eq. (4.229) in the last step,

$$\begin{aligned} |f_\omega(Q), g_\omega(Q)| &= |p_0(Q) - p_\omega, \; q_0(Q) - Q| \\ &= |(p_0(Q), q_0(Q)) - (p_\omega, Q)| \\ &= |(\tilde{p}_0(t), \tilde{q}_0(t)) - (p_\omega, q_\infty(t))| \\ &= |S_\infty(p_\infty, q_\infty(t)) - (p_\omega, q_\infty(t))| < \kappa, \end{aligned} \tag{4.235}$$

which is Eq. (4.187) in conclusion (III) of the KAM theorem.

We show that f_ω and g_ω are *analytic*, as stated in conclusion (II) of the KAM theorem. According to Eq. (4.234), $f_\omega(Q)$ and $g_\omega(Q)$ are analytic if $p_0(Q)$ and $q_0(Q)$ are. The latter are given by Eq. (4.233), thus S_∞ must be shown to be analytic with respect to its second argument Q. Each of the canonical transformations B_s is a diffeomorphism, hence analytic, thus $S_s = B_1 \circ \cdots \circ B_s$ is analytic. According to Lemma C1, the S_s converge uniformly to S_∞ on F_∞ for each fixed $p_\infty \in G_\infty$. The q-domain is Re $q_i \in [0, 2\pi]$ and Im $q_i \in [-\rho_\infty, \rho_\infty]$, i.e., compact. From the Weierstrass theorem on uniformly convergent sequences of analytic functions,[47] it follows that S_∞ is analytic, and so are f_ω and g_ω.

The *motion* on T_ω is given by $Q = q_\infty(t) = \omega t = A_0(p_\omega)t$, which corresponds to Eq. (4.188) from conclusion (IV) of the KAM theorem.

The *invariance* of T_ω (i.e., the trajectory stays on the torus for all times) stated in conclusion (II) was shown in (iv). This fully proves the KAM theorem. $\qquad\square$

---

[47] See Markushevich (1965), p. 330, or Ahlfors (1966), p. 174. We cite the theorem from the latter: 'Suppose that $f_n(z)$ is analytic in the region Ω_n, and that the sequence $\{f_n(z)\}$ converges to a limit function $f(z)$ in a region Ω, uniformly on every compact subset of Ω. Then $f(z)$ is analytic in Ω.'

5 Analytic Lemmas

The present and the next three chapters state and prove 20 lemmas used in the KAM proof of the foregoing chapter. Lemmas G1, G5 and D3 are important theorems in themselves (G1 is equivalent to an important lemma by Brouwer 1911a). The following four 'analytic lemmas' numbered A1 to A4 are contained in section 4.2 of Arnold (1963a) and give limits for the Fourier coefficients and the derivatives of analytic functions.

5.1 LEMMA A1. EXPONENTIALS VS. POWERS

Lemma A1 *For any $m > 0, v > 0$ and $\delta > 0$,*

$$m^v \leq \left(\frac{v}{e}\right)^v \frac{e^{m\delta}}{\delta^v}, \tag{5.1}$$

$$\ln \frac{1}{\delta} \leq \frac{v}{e} \left(\frac{1}{\delta}\right)^{\frac{1}{v}}. \tag{5.2}$$

Proof Consider the function $f(x)$ and its first two derivatives for $x > 0$,

$$f(x) = x - v \ln x,$$
$$f'(x) = 1 - \frac{v}{x},$$
$$f''(x) = \frac{v}{x^2} > 0. \tag{5.3}$$

Thus $x = v$ is a minimum, and $f(x) \geq f(v)$ for all $x > 0$, or

$$x - v \ln x \geq v - v \ln v,$$
$$\frac{e^x}{x^v} \geq \frac{e^v}{v^v}. \tag{5.4}$$

Let $x = m\delta$. Then Eq. (5.4) gives

$$\frac{e^{m\delta}}{(m\delta)^v} \geq \frac{e^v}{v^v} \tag{5.5}$$

or

$$m^v \leq \left(\frac{v}{e}\right)^v \frac{e^{m\delta}}{\delta^v}, \tag{5.6}$$

DOI: 10.1201/9781003287803-5

which is Eq. (5.1). Next, for $0 < \delta < 1$, let $x = \ln \dfrac{1}{\delta} > 0$. Then, again from Eq. (5.4),

$$\frac{e^{\ln \frac{1}{\delta}}}{\left(\ln \frac{1}{\delta}\right)^\nu} \geq \frac{e^\nu}{\nu^\nu} \tag{5.7}$$

or

$$\ln \frac{1}{\delta} \leq \frac{\nu}{e} \left(\frac{1}{\delta}\right)^{\frac{1}{\nu}}, \tag{5.8}$$

which is Eq. (5.2). The case $\delta > 1$ is done similarly but is not needed in the following.
□

Remark The purpose of Eq. (5.1) is to replace a power x^n (with $x = m\delta$ here) by an exponential e^x. Since the $e^0 > 0^n$ and $e^x > x^n$ for any n for sufficiently large x, it is clear that there is a factor (determined here as $C(n) = (n/e)^n$) so that $C(n)e^x > x^n$ for all $x > 0$. The estimate (5.1) is used in Eq. (5.31) of Lemma A2 below and Eq. (4.61) of the Fundamental theorem; the estimate (5.2) is used in Eq. (4.147) of the Inductive theorem. The function $x - n \ln x$ used above is shown in Fig. 5.1.

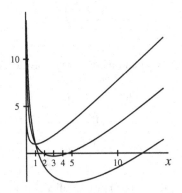

Figure 5.1   The function $x - n \ln x$ for $n = 1, 3, 5$

5.2   FUNCTIONS IN SEVERAL COMPLEX VARIABLES

In the KAM theorem complex functions H, A, S, B with several complex variables occur. We list here some definitions and theorems for such functions, taken from Rothstein & Kopfermann (1982). The central result is that the Cauchy integral formula applies for n complex variables, with n loop integrals in n distinct complex planes. In this section we use the generic variable z from complex function theory, and return to the KAM variable x (especially $x = (p, q)$) in the next section.

Definition A *domain* $U \subset \mathbb{C}^n$ is an open, connected set. Here 'open' refers to the topology induced by the metric with distance function $d(z,z') = |z - z'|$, which in turn is induced by the maximum norm $|z| = \max(|z_1|, \ldots, |z_n|)$.

Definition A complex function $f : G \subset \mathbb{C}^n \to \mathbb{C}$ with open G is *analytic in $a \in G$* if there is an open neighborhood $U \subset G$ of a and a power series

$$P(z) = \sum_\nu b_\nu (z-a)^\nu \tag{5.9}$$

convergent in U and such that $f(z) = P(z)$ for all $z \in U$. Here $\nu = (\nu_1, \ldots, \nu_n) \in \mathbb{N}^n$ (including zero) is a multi-index, $b_\nu = b_{\nu_1, \ldots, \nu_n} \in \mathbb{C}$, and $z^\nu = z_1^{\nu_1} \ldots z_n^{\nu_n}$. The function f is *analytic in G* if it is analytic in each $a \in G$; and f is *analytic in $M \subset G$* (M open, closed, or neither) if there is an open set N such that $M \subset N \subset G$ and f is analytic in N.

Definition A function $f(z_1, \ldots, z_n)$ is *complex partial differentiable in a* if df/dz_i exists in a for each z_i in the usual sense of complex function theory, i.e., with $z_j = a_j$ held fixed for $j \neq i$. f is *partial differentiable in G* if it is so in all $a \in G$. One writes $\partial_i f = df/dz_i$ with all $z_j, j \neq i$ held fixed.

Definition A function $f : G \subset \mathbb{C}^n \to \mathbb{C}$ with G a domain is *complex differentiable in $a \in G$* if there are n functions $g_i : G \to \mathbb{C}$ continuous in a such that for all $z \in G$,

$$f(z) = f(a) + \sum_{i=1}^{n} (z_i - a_i) g_i. \tag{5.10}$$

If f is complex differentiable in all $a \in G$, it is *complex differentiable in G*.

Lemma A complex differentiable function f is continuous and complex partially differentiable.

Proof $g_i = \partial_i f$.

Lemma An analytic function is complex differentiable.

Proof Differentiation of its power series.

Definition Let $G_1, \ldots, G_n \subset \mathbb{C}$ be domains. Then a *polydomain G* is the topological product

$$G = G_1 \times \cdots \times G_n. \tag{5.11}$$

Let $D_1, \ldots, D_n \subset \mathbb{C}$ with $D_i = \{z_i \in \mathbb{C} : |z_i - a_i| < r_i\}$ be open disks, where each $r_i \in \mathbb{R}$ can be finite or infinite. A *polydisk D* is[1] the topological product

$$D = D_1 \times \cdots \times D_n. \tag{5.12}$$

---

[1] D is not to be confused with the ratio of surface to volume introduced in Section 8.1.

D has complex dimension n and real dimension $2n$, the boundary ∂D has real dimension $2n - 1$. Of central importance is the *distinguished boundary*

$$\partial_0 D = \partial D_1 \times \cdots \times \partial D_n. \tag{5.13}$$

Since each ∂D_i is a circle, $\partial_0 D \subset \partial D$ is a torus of real dimension n.

Theorem (Cauchy integral formula) Let $f : G \subset \mathbb{C}^n \to \mathbb{C}$ be continuous and complex partial differentiable in G. Then for any point z of a polydisk $D \subset G$,

$$
\begin{aligned}
f(z) &= \frac{1}{(2\pi i)^n} \int_{\partial_0 D} d\zeta_1 \ldots d\zeta_n \frac{f(\zeta)}{(\zeta_1 - z_1) \ldots (\zeta_n - z_n)} \\
&= \frac{1}{2\pi i} \oint_{\partial D_1} \frac{d\zeta_1}{\zeta_1 - z_1} \cdots \frac{1}{2\pi i} \oint_{\partial D_n} \frac{d\zeta_1}{\zeta_n - z_n} f(\zeta_1, \ldots, \zeta_n).
\end{aligned}
\tag{5.14}
$$

Thus the values of f on the n circles (or straight infinite lines) ∂D_i is sufficient to determine f in the polydisk $D = D_1 \times \cdots D_n$. This formula also holds if D is replaced by a polydomain $G = G_1 \times \cdots \times G_n$ where all boundary lines ∂G_i are piecewise smooth.

Proof Apply Cauchy's integral formula from complex function theory of one variable to z_1, then to z_2, up to z_n.

Theorem (Cauchy integral formula for partial derivatives) Let $v = (v_1, \ldots, v_n) \in \mathbb{N}^n$ and $f : G \subset \mathbb{C}^n \to \mathbb{C}$ be continuous and complex partial differentiable in G. Then in any polydisk $D \subset G$, all partial derivatives of f exist, and for any $z \in D$,

$$\frac{\partial^{v_1 + \cdots + v_n} f(z)}{\partial z_1^{v_1} \ldots \partial z_n^{v_n}} = \frac{v_1! \ldots v_n!}{(2\pi i)^n} \int_{\partial_0 D} d\zeta_1 \ldots d\zeta_n \frac{f(\zeta_1, \ldots, \zeta_n)}{(\zeta_1 - z_1)^{v_1+1} \ldots (\zeta_n - z_n)^{v_n+1}}. \tag{5.15}$$

Proof Differentiate the Cauchy integral formula and exchange differentiation and integration.

Theorem (Osgood) If f is continuous and complex partial differentiable in a domain $G \subset \mathbb{C}^n$, then it is complex differentiable in G, hence analytic.

5.3   LEMMA A2. FOURIER COEFFICIENTS

A periodic, integrable function f is analytic (complex differentiable) if and only if its Fourier coefficients f_k decay exponentially with k, $|f_k| \le C e^{-c|k|}$ with constants $c, C > 0$. The following lemma relates bounds on analytic functions and their Fourier coefficients.

Lemma A2 *Let* $f(q) = f(q_1, \ldots, q_n)$ *be analytic in the complex strips* [2] $\{q_i \in \mathbb{C} :$ $|\mathrm{Im}\, q_i| \leq \rho\}$. *Let* $f(q)$ *be* 2π-*periodic in* $\mathrm{Re}\, q_1$ *to* $\mathrm{Re}\, q_n$, *with Fourier series*

$$f(q) = \sum_{k_1 = -\infty}^{\infty} e^{ik_1 q_1} \cdots \sum_{k_n = -\infty}^{\infty} e^{ik_n q_n}\, f_{k_1 \ldots k_n} = \sum_k f_k e^{i(k,q)}, \qquad (5.16)$$

where $k = (k_1, \ldots, k_n)$ *with* $k_i \in \mathbb{Z}$. *Let* (*with 'R' for remainder*)

$$R_N f(q) = \sum_{|k| \geq N} f_k e^{i(k,q)}. \qquad (5.17)$$

Then the following bounds hold.

(I) *For* $\rho > 0$:

$$\left\{ \text{If} \quad |f(q)| \leq M \quad \text{for} \quad |\mathrm{Im}\, q| \leq \rho \right\} \quad \text{then} \quad |f_k| \leq M e^{-|k|\rho}. \qquad (5.18)$$

(II) *For* [3] $0 < \delta < \rho \leq 1$:

$$\text{If} \quad |f_k| \leq M e^{-|k|\rho} \quad \text{then} \quad \left\{ |f(q)| < 4^n \frac{M}{\delta^n} \quad \text{for} \quad |\mathrm{Im}\, q| \leq \rho - \delta \right\}. \qquad (5.19)$$

(III) *For* $2\delta < \gamma$ *and* $\delta + \gamma < \rho \leq 1$:

$$\text{If } |f_k| \leq M e^{-|k|\rho} \text{ then } \left\{ |R_N f(q)| < \frac{(2n)^n}{e^n} \frac{M}{\delta^{n+1}} e^{-N\gamma} \text{ for } |\mathrm{Im}\, q| \leq \rho - \delta - \gamma \right\}. \qquad (5.20)$$

Remark The bounds for $|f(q)|$ and $|R_N f(q)|$ in (II) and (III) are direct consequences of the domain restrictions of $|\mathrm{Im}\, q|$ and not possible without the latter. A major effort in the proof of the KAM theorem is to show that iterated domain reductions leave non-empty domains (and even large in terms of their measure).

Proof of (I) The Fourier coefficients are given by

$$f_{k_1 \ldots k_n} = \frac{1}{2\pi} \int_0^{2\pi} dq_1\, e^{-ik_1 q_1} \cdots \frac{1}{2\pi} \int_0^{2\pi} dq_n\, e^{-ik_n q_n}\, f(q), \qquad (5.21)$$

or simply

$$f_k = \frac{1}{(2\pi)^n} \int dq\, f(q)\, e^{-i(k,q)} \qquad (5.22)$$

---

[2] See Section 2.1 for analytic functions on closed domains. f has to be analytic on an open domain that contains the closed domain.

[3] The $f(q)$ in statements (I) and (II) are general different functions, as are the f_k. Similarly, ρ in (I) and (II) are not (directly) related. Statements (II) and (III), by contrast, refer to the same function f.

where $dq = dq_1 \ldots dq_n$. Now

$$\left| e^{-i(k,q)} \right| = \left| e^{-i(k,\mathrm{Re}(q))} \right| e^{(k,\mathrm{Im}(q))} = e^{\Sigma_i k_i \, \mathrm{Im}(q_i)}. \tag{5.23}$$

If $k_i < 0$, choose $\mathrm{Im}(q_i) = +\rho$, and if $k_i > 0$, choose $\mathrm{Im}(q_i) = -\rho$. Then

$$\left| e^{-i(k,q)} \right| = e^{-\rho \Sigma_i |k_i|} = e^{-|k|\rho}. \tag{5.24}$$

Consider the integral over any of the q_i and assume $k_i > 0$. Since f is analytic in the simply connected domain of the complex q_i-plane shown in Fig. 5.2, the integral along the loop C in the lower complex half-plane is zero according to Cauchy's theorem. Since f is 2π-periodic, the contributions from the segments B and B' cancel each other. Thus the integral from 0 to 2π along the real axis equals the integral along the lower straight line A. For $k_i < 0$, integrate instead along A'.[4]

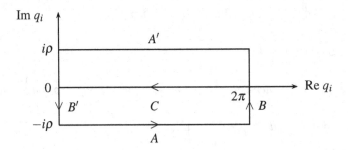

Figure 5.2 Periodic complex function analytic in a ρ-strip

One has then

$$|f_k| = \frac{1}{(2\pi)^n} \left| \int dq \, f(q) \, e^{-i(k,q)} \right|$$

$$\leq \frac{1}{(2\pi)^n} \int dq \, |f(q)| \, \left| e^{-i(k,q)} \right|$$

$$\leq \frac{M e^{-|k|\rho}}{(2\pi)^n} \int_0^{2\pi} dq_1 \ldots \int_0^{2\pi} dq_n$$

$$= M e^{-|k|\rho}. \tag{5.25}$$

Proof of (II)   As in Eq. (5.23),

$$\left| e^{i(k,q)} \right| = e^{-\Sigma_i k_i \, \mathrm{Im}(q_i)}, \tag{5.26}$$

---

[4]Arnold (1963a) states that the contour of integration is shifted to $i\rho$, Arnold (1963b) that it is shifted to $\pm i\rho$.

where q_i is restricted to the strip $|\mathrm{Im}\, q_i| \leq \rho - \delta$. The function $f(q)$ is given by a Fourier series over all k, including the exponential term from Eq. (5.26). For any given q_i with positive or negative imaginary part, both $+k_i$ and $-k_i$ appear in the Fourier series; one of the latter two gives an exponential $\exp[(\rho - \delta)|k_i|]$ with positive argument $(\delta < \rho)$, entering the bound for $f(q)$. Collecting the contributions from k_1 to k_n to the exponential gives

$$\left| e^{i(k,q)} \right| \leq e^{(\rho-\delta)\Sigma_i |k_i|} = e^{|k|(\rho-\delta)} \qquad \text{for} \qquad |\mathrm{Im}\, q| \leq \rho - \delta. \qquad (5.27)$$

With the abbreviation $y = e^{-\delta}$ and the sum formula for the geometric series,

$$\sum_{m=0}^{\infty} y^m = \frac{1}{1-y}, \qquad (5.28)$$

one obtains

$$
\begin{aligned}
|f(q)| &\leq \sum_k |f_k| \left| e^{i(k,q)} \right| \\
&\leq M \sum_k e^{-|k|\rho}\, e^{|k|(\rho-\delta)} \\
&= M \sum_{k_1=-\infty}^{\infty} \cdots \sum_{k_n=-\infty}^{\infty} e^{-(|k_1|+\cdots+|k_n|)\delta} \\
&= M \left(\sum_{k_1=-\infty}^{\infty} e^{-|k_1|\delta} \right)^n \\
&= M \left(1 + 2 \sum_{m=1}^{\infty} e^{-m\delta} \right)^n \\
&= M \left(1 + 2y \sum_{m=0}^{\infty} y^m \right)^n \\
&= M \left(1 + \frac{2e^{-\delta}}{1-e^{-\delta}} \right)^n \\
&= M \left(\frac{1+e^{-\delta}}{1-e^{-\delta}} \right)^n \\
&< M \left(\frac{4}{\delta} \right)^n, \qquad (5.29)
\end{aligned}
$$

using the formula

$$\frac{1+e^{-\delta}}{1-e^{-\delta}} < \frac{4}{\delta} \qquad \text{for} \qquad 0 < \delta \leq 1, \qquad (5.30)$$

which is easily checked.

Proof of (III)   For q restricted to $|\mathrm{Im}\, q| \leq \rho - \delta - \gamma$ one has, replacing v by n in Eq. (5.1),

$$
\begin{aligned}
|R_N f(q)| &= \sum_{|k| \geq N} |f_k| \left| e^{i(k,q)} \right| \\
&\overset{(5.27)}{\leq} \sum M e^{-|k|\rho}\, e^{|k|(\rho-\delta-\gamma)} \\
&\overset{(8.19)}{\leq} M \sum_{m \geq N} 2^n m^{n-1}\, e^{-m(\delta+\gamma)} \\
&< M \sum_{m \geq N} 2^n m^n\, e^{-m\delta}\, e^{-m\gamma} \\
&\overset{(5.1)}{\leq} M \sum_{m \geq N} 2^n m^n \left(\frac{n}{e} \right)^n \frac{1}{m^n \delta^n}\, e^{-m\gamma} \\
&= \left(\frac{2n}{e} \right)^n \frac{M}{\delta^n} \sum_{m \geq N} e^{-m\gamma} \\
&= \left(\frac{2n}{e} \right)^n \frac{M}{\delta^n}\, e^{-N\gamma} \sum_{m \geq 0} e^{-m\gamma} \\
&= \left(\frac{2n}{e} \right)^n \frac{M}{\delta^n}\, \frac{e^{-N\gamma}}{1-e^{-\gamma}} \\
&< \left(\frac{2n}{e} \right)^n \frac{M}{\delta^{n+1}}\, e^{-N\gamma},
\end{aligned}
\tag{5.31}
$$

using the elementary estimate

$$
1 - e^{-\gamma} > \delta \qquad \text{for} \qquad 2\delta < \gamma < 1,
\tag{5.32}
$$

which is easily checked; and $2\delta < \gamma < 1$ is true by assumption of the lemma.   □

5.4   LEMMA A3. CAUCHY ESTIMATES

The following important estimates use Cauchy's integral formula. In the foregoing lemma, we considered complex variables q_i that were periodic in their real parts. Here, x_i are general complex variables (the lemma is applied to the momentum variable).

Lemma A3 *Let* $f(x) = f(x_1, \ldots, x_n)$ *be analytic on the complex domain* $U \subset \mathbb{C}^n$, *with* $|f(x)| \leq M$ *for all* $x \in U$. *Then on the smaller domain* $U - \delta$, *for* $i, j = 1, \ldots, n$,

$$
\left\| \frac{\partial f}{\partial x_i} \right\| \leq \frac{M}{\delta}, \qquad\qquad \left\| \frac{\partial^2 f}{\partial x_i \partial x_j} \right\| \leq \frac{2M}{\delta^2}.
\tag{5.33}
$$

Remark If $x = q$ is again an angle variable defined in complex strips $0 \leq \mathrm{Re}\, q_i \leq 2\pi$ and $\mathrm{Im}\, q_i \leq \rho$, then for $\|f\|_\delta = \sup_q |f(q)|$ with q from these strips, de la Llave

(2001), p. 30, writes the first inequality in (5.33) more suggestively as

$$\left\| \frac{\partial f}{\partial x_i} \right\|_{\rho-\delta} \le \frac{\|f\|_\rho}{\delta}. \tag{5.34}$$

Proof The Cauchy formula for the n-th derivative of an analytic function of a single complex variable is

$$\frac{\mathrm{d}^n f(x)}{\mathrm{d}x^n} = \frac{n!}{2\pi i} \oint_C \mathrm{d}\xi \, \frac{f(\xi)}{(\xi - x)^{n+1}}, \tag{5.35}$$

which for several complex arguments becomes Eq. (5.15),

$$\frac{\partial^{v_1 + \cdots + v_n} f(x)}{\partial x_1^{v_1} \dots \partial x_n^{v_n}} = v_1! \dots v_n! \frac{1}{2\pi i} \oint_{C_1} \mathrm{d}\xi_1 \dots \frac{1}{2\pi i} \oint_{C_n} \mathrm{d}\xi_n \frac{f(\xi_1, \dots, \xi_n)}{(\xi_1 - x_1)^{v_1+1} \dots (\xi_n - x_n)^{v_n+1}},$$

with $v_1 + \cdots + v_2 = n$ the degree of differentiation and closed loops $C_1, \dots, C_n$ lying fully within the domain of analyticity. Let $C_1, \dots, C_n$ be circles with centers $\xi_1, \dots, \xi_n$ in the respective complex planes. By restricting all the ξ_i to $U - \delta$, the maximum allowed radius for each circle near the boundary $\partial(U - \delta)$ is δ in order not to leave U. Thus, using polar representations $\xi_i = \delta e^{i\varphi_i} + x_i$,

$$\left\| \frac{\partial^{v_1 + \cdots + v_n} f}{\partial x_1^{v_1} \dots \partial x_n^{v_n}} \right\| \le v_1! \dots v_n! \, M \frac{1}{2\pi} \int_0^{2\pi} \frac{\delta \, \mathrm{d}\varphi_1}{\delta^{v_1+1}} \dots \frac{1}{2\pi} \int_0^{2\pi} \frac{\delta \, \mathrm{d}\varphi_n}{\delta^{v_n+1}}, \tag{5.36}$$

or

$$\left\| \frac{\partial^{v_1 + \cdots + v_n} f}{\partial x_1^{v_1} \dots \partial x_n^{v_n}} \right\| \le \frac{v_1! \dots v_n!}{\delta^{v_1 + \cdots + v_n}} \, M. \tag{5.37}$$

For $n = v_i = 1$ and $v_j = 0$ for $j \ne i$, one has $(i, j = 1, \dots, n)$

$$\left\| \frac{\partial f}{\partial x_i} \right\| \le \frac{M}{\delta}, \tag{5.38}$$

which is the first inequality in Eq. (5.33). For $n = v_i = 2$ and $v_j = 0$ for $j \ne i$,

$$\left\| \frac{\partial^2 f}{\partial x_i^2} \right\| \le \frac{2M}{\delta^2}, \tag{5.39}$$

while for $n = 2$ and $v_i = v_j = 1$ with $i \ne j$ and $v_k = 0$ for $j \ne k \ne i$,

$$\left\| \frac{\partial^2 f}{\partial x_i \partial x_j} \right\| \le \frac{M}{\delta^2}. \tag{5.40}$$

The second inequality in (5.33) covers both (5.39) and (5.40). $\qquad \square$

The generalization of this lemma to functions of several complex variables is obvious.

5.5   LEMMA A4. LAGRANGE AND TAYLOR ESTIMATES

The following lemma uses the Lagrange and Taylor formulas.

Lemma A4   *Let $f = (f_1, \ldots, f_m)$ be a vector-valued, continuously differentiable function in $\mathbb{R}^n$ with points $x = (x_1, \ldots, x_n)$.*
(I) If for all points x on the straight segment $\overline{ab}$,

$$|df(x)| \le C|dx| \tag{5.41}$$

with C some constant, then

$$|f(b) - f(a)| \le C|b - a|. \tag{5.42}$$

(II) If for all x along $\overline{ab}$ and for all $i = 1, \ldots, m$ and $j = 1, \ldots, n$

$$\left| \frac{\partial f_i}{\partial x_j} \right| \le c, \tag{5.43}$$

then

$$|f(b) - f(a)| \le cn|b - a|. \tag{5.44}$$

(III) If f is twice continuously differentiable on $\overline{ab}$ and

$$\left| \frac{\partial^2 f_i}{\partial x_j \, \partial x_k} \right| \le \Theta \tag{5.45}$$

for all i, j, k with constant Θ, then

$$\left| f(b) - f(a) - \left(\frac{\partial f}{\partial x} \bigg|_a, b - a \right) \right| \le \frac{\Theta n^2}{2} |b - a|^2. \tag{5.46}$$

Proof of (I)

$$|f(b) - f(a)| = \left| \int_{f(a)}^{f(b)} df \right| \le \int_{f(a)}^{f(b)} |df| \le C \int_a^b |dx| \overset{(\P)}{=} C \left| \int_a^b dx \right| = C|b - a|. \tag{5.47}$$

In ($\P$) we used that integration is along a straight segment from a to b.

Proof of (II)

$$|f(b) - f(a)| = \max\left\{\left|\int_a^b \sum_{j=1}^n \frac{\partial f_1}{\partial x_j}\, dx_j\right|, \ldots, \left|\int_a^b \sum_{j=1}^n \frac{\partial f_m}{\partial x_j}\, dx_j\right|\right\}$$

$$\leq \max\left\{\int_a^b \sum_{j=1}^n \left|\frac{\partial f_1}{\partial x_j}\right| |dx_j|, \ldots, \int_a^b \sum_{j=1}^n \left|\frac{\partial f_m}{\partial x_j}\right| |dx_j|\right\}$$

$$\leq c \sum_{j=1}^n \int_{a_j}^{b_j} |dx_j|$$

$$= c \sum_{j=1}^n \left|\int_{a_j}^{b_j} dx_j\right|$$

$$\leq cn \max_j\{|b_j - a_j|\}$$

$$= cn\, |b-a|, \tag{5.48}$$

where again integration along a straight line was used.

Proof of (III)

$$\left|f(b) - f(a) - \left(\frac{\partial f}{\partial x}\bigg|_a, b - a\right)\right|$$

$$= \max_i\left\{\left|f_i(b) - f_i(a) - \sum_{j=1}^n \frac{\partial f_i}{\partial x_j}\bigg|_a (b-a)_j\right|\right\}$$

$$= \max_i\left\{\left|\int_a^b \sum_{j=1}^n \left(\frac{\partial f_i(x)}{\partial x_j} - \frac{\partial f_i}{\partial x_j}\bigg|_a\right) dx_j\right|\right\}$$

$$\overset{(a)}{=} \max_i\left\{\left|\int_a^b \sum_{j=1}^n \left(\int_a^x \sum_{k=1}^n \frac{\partial^2 f_i}{\partial x_j\, \partial x_k}\, dx_k\right) dx_j\right|\right\}$$

$$\leq \Theta \sum_{j=1}^n \sum_{k=1}^n \int_{a_j}^{b_j} \int_{a_k}^{x_k} |dx_k\, dx_j|$$

$$\leq \Theta n^2 \max_{k,j} \int_{a_j}^{b_j} \int_{a_k}^{x_k} |dx_k\, dx_j|$$

$$\overset{(b)}{=} \Theta n^2 \int_{a_m}^{b_m} \int_{a_m}^{x_m} |dx_m\, dx_m'|$$

$$\overset{(c)}{=} \frac{\Theta n^2}{2} \int_{a_m}^{b_m} |dx_m| \left(\int_{a_m}^{b_m} |dx_m'|\right)$$

$$\overset{(d)}{=} \frac{\Theta n^2}{2} |b-a|^2. \tag{5.49}$$

In (a), $f'(x) - f'(a)$ is replaced by $\int_a^x dx\, f''$ (for an n-tuple x_i); note the upper limit in this integral. In (b), m is the index for which the double integral is maximal. Since $\max_{j,k}\{|x_j x_k|\} = (\max_j\{|x_j|\})^2$ for a tuple (x_j), we can assume $j = k = m$ for some m.

The prime at the second x_m indicates that two separate integrations are performed. In (c), a standard integration formula is used, see Fig. 5.3. In (d), we use again that $|b_m - a_m| = \max_{j}\{|b_j - a_j|\}$. $\qquad\qquad$ □

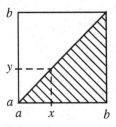

Figure 5.3 $\int_a^b \left(\int_a^x dy\right) dx = \frac{1}{2}\int_a^b \int_a^b dx\, dy$

6 Geometric Lemmas

We turn to the important section 4.3 'Geometric lemmas' in Arnold (1963a) containing five lemmas, numbered G1 to G5 in the following. They mostly address the diffeomorphism A from momentum to frequency space, which is derived from the integrable part $H^+(p)$ of the Hamiltonian. The change in H^+ due to a perturbation causes a change of A to A', i.e., a shift in torus frequencies. The perturbed frequencies have to lie sufficiently close to the unperturbed frequencies, so that if the latter obey a Diophantine condition, the perturbed frequencies obey a slightly weaker Diophantine condition too.

Lemma G5 is pivotal to the KAM proof and defines the new domain on which A' is a diffeomorphism. The proof of Lemma G5 uses Lemmas G1, G2, G4.

Lemma G1 uses an important theorem from algebraic topology on the homotopy invariance of the mapping degree, which is one of the few known topological invariants. The concept of the mapping degree will be developed in an elementary way, starting with curves.

Since the proof of Lemma G5 is somewhat intricate, we prove in the next section a lemma that catches the essentials of Lemma G5 in a simpler setting. The proof of Lemmas G1 to G5 will not refer directly to this Section 6.1, which can therefore also be skipped.

6.1 AN IMPLICIT FUNCTION THEOREM

Lemma G5 is an inverse function theorem, establishing the existence of a diffeomorphism A' (thus A' is invertible) for a perturbed diffeomorphism A. The reduction in domain size of A' is directly related to the deviation of A' from A. The present section treats such a situation in a rather simple inverse function theorem taken from Pöschel (2001), termed an implicit function theorem there. The proof of an inverse function theorem from an implicit function theorem is standard (or 'trivial,' see Krantz & Parks 2013, p. 44). The opposite direction is also not too hard, see again there. KAM theorems can be formulated (to some degree) as implicit function theorems:

> This cumbersome iterative procedure [infinite sequence of canonical transformations] is similar to Newton's method of tangents for solving algebraic equations. There are also approaches using 'hard' Implicit Function Theorems in infinite-dimensional spaces. (Broer & Sevryuk 2010, pp. 260-261)

Or:

> [...] some of the abstract versions of KAM as an implicit function theorem work perfectly well for Sobolev spaces. I think it is mainly a historical anomaly

DOI: 10.1201/9781003287803-6

that these spaces are not used more frequently in the KAM theory of dynamical systems. (de la Llave 2001, p. 27)

For more information on this approach, see, e.g., chapter 7 in Sternberg (1969), 'The KAM implicit function theorem,' chapter 2 in Schwartz (1969), 'Hard implicit functional theorems' and two papers by Zehnder (1975, 1976). A book starting with the historical development of the implicit function theorem and ending with the Nash-Moser theorem is Krantz & Parks (2013). The excellent sections 3.2 to 3.4 of this book contain three different detailed proofs of the implicit function theorem (induction; standard analysis; Banach contraction principle). This latter (standard) proof can also be found in appendix I of Sternberg (1964) or in Forster (1979), pp. 66-76. Some of Arnold's arguments presented in the next lemmas come quite close to those used by Krantz & Parks (2013), so that a study of this reference is helpful for the following material.

Let $G \subset \mathbb{R}^n$ be a domain. To make theorems for analytic functions available (especially Cauchy's theorem), G is made subset of a complex domain G_h,

$$G_h = \{x \in \mathbb{C}^n : \forall y \in G \ |x - y| < h\}, \tag{6.1}$$

i.e., imaginary directions are added to G. Let

$$\|f\|_G = \sup_{x \in G} |f(x)|. \tag{6.2}$$

For the next lemma and its proof, see Pöschel (2001), p. 30. (For a simple inverse function theorem, see also lemma 2.3 in Wayne 1996, p. 9.)

Lemma *Let $f : G_h \to \mathbb{C}^n$ be analytic with*

$$\|f - E\|_{G_h} \leq h/4, \tag{6.3}$$

where E is the identity map. Then f has an analytic inverse f^{-1} on the domain $G_{h/4}$ and

$$\left\|f^{-1} - E\right\|_{G_{h/4}} \leq h/4. \tag{6.4}$$

Proof We show that f is injective and surjective on $G_{h/4}$. Regarding injectivity, assume $f(u) = f(v)$ for $u, v \in G_{h/2}$. Then

$$|u - v| = |u - f(u) - v + f(v)| \leq |u - f(u)| + |v - f(v)| \overset{(6.3)}{\leq} h/2. \tag{6.5}$$

Consider the straight segment $(1 - s)u + sv$ with $0 \leq s \leq 1$ that connects u and v. We show that this segment lies in $G_{3h/4}$. Indeed, let $b \in \partial G_h$ be a boundary point of G_h and q be any point on $\overline{uv}$ for $0 \leq s \leq 1/2$. Then

$$|q - b| = |(1 - s)u + sv - b| = |(1 - s)u + sv - u + (u - b)|$$

$$> \Big| |u - b| - s|u - v| \Big| \geq \frac{h}{2} - \frac{1}{2}\frac{h}{2} = \frac{h}{4}, \tag{6.6}$$

where $|u - b| \geq h/2$ since $u \in G_{h/2}$. From $|q - b| \geq h/4$ follows $q \in G_{3h/4}$. Similarly, for $1/2 < s \leq 1$ put $t = 1 - s$ and consider $|q - b| = |tu - (1 - t)v - v + (v - b)|$, giving again $|q - b| > h/4$.

Since the boundary of G_h is at a distance $> h/4$ along the whole segment $\overline{uv}$, the Cauchy estimate (5.33) for the first derivative gives along $\overline{uv}$ (with $\mathbb{I}$ the unit matrix),

$$\eta = \max_{\overline{uv}} \ |Df - \mathbb{I}| = \max_{\overline{uv}} \ |D(f - E)| < \frac{h/4}{h/4} = 1. \tag{6.7}$$

According to the mean value theorem, there is a $q \in \overline{uv}$ such that

$$D(f - E)(q) = \frac{(f - E)(v) - (f - E)(u)}{v - u} \tag{6.8}$$

and thus, interchanging the two sides of the equation and since $f(u) = f(v) = 0$,

$$1 = \frac{|u - v|}{|v - u|} = |D(f - E)(q)| \leq \eta < 1, \tag{6.9}$$

which is a contradiction unless $u = v$. Thus f is one-to-one on $G_{h/2}$.

Pöschel (2001), p. 31, states that 'by elementary arguments from degree theory the image of $G_{h/2}$ under f covers $G_{h/4}$, since $\|f - E\| \leq h/4$' (symbols adapted). We will discuss these arguments from degree theory in the next sections. The arguments may be elementary, but they are not obvious: although f can shift points by distances $\leq h/4$ only, one can still imagine that the image $f(G_{h/2})$ has a 'hole' in $G_{h/4}$, since continuous mappings of simply connected domains need not have simply connected images. Having shown that f is injective on $G_{h/4}$ and assuming that it is also surjective on $G_{h/4}$, f has an inverse map $f^{-1} : G_{h/4} \to \mathbb{C}^n$. For the preimage $f^{-1}(v)$ of a point $v = f(u)$ of $G_{h/4}$,

$$|f^{-1}(v) - v| = |f^{-1}(f(u)) - f(u)| = |u - f(u)| \leq h/4, \tag{6.10}$$

thus $\|f^{-1} - E\| \leq h/4$ on $G_{h/4}$, which proves Eq. (6.4). □

6.2 MAPPING DEGREE

Brouwer's approach to degree was based on his fundamental idea that given a map $f : M \to N$ between n-dimensional manifolds, much information about the map can be obtained from the study of the inverse images of a single point in N. He defined the degree of f (by first taking a simplicial approximation φ of f) as 'number of inverse images of a point' (counted with appropriate multiplicities) and then proved that the degree is a homotopy invariant of f. (Granas & Dugundji 2003, p. 276)

The function $f : \mathbb{R} \to \mathbb{R}_+, x \mapsto x^2$, for example, has degree zero: each $y > 0$ has two preimages (inverse images), $\pm \sqrt{y}$. For $x < 0$, $y' < 0$ and the multiplicity is counted as -1 (since $dx > 0$ implies $dy < 0$), while $y' > 0$ for $x > 0$ and the multiplicity is $+1$. The degree is then $d = 1 - 1 = 0$ for $y > 0$; and correspondingly for $y < 0$, $d = 0 - 0 = 0$.

Two continuous functions $f, g : X \to Y$ are *homotopic* if there is a continuous map $H : X \times [0,1] \to Y$ with $H(x,0) = f(x)$ and $H(x,1) = g(x)$ for all $x \in X$. Note that the image of H must always lie in Y. Thus the identity on the unit circle, $E : S \to S, x \mapsto x$ in the Euclidean plane is not homotopic to the constant map, $0 : S \to 1$, since 'contracting' S to a point leads through the disk $r < 1$ not belonging to S.

We start with the definition of the *index* or *winding number* of a closed curve C about a point p in the Euclidean plane. Let $C : [\xi_0, \xi_1] \ni \xi \mapsto C(\xi) \in \mathbb{R}^2$ be an arbitrary smooth curve in $\mathbb{R}^2$ and a and x be a fixed and variable point on C, respectively, and p a point not lying on C. Let $F_a(p, C, x)$ be the angle centered at p between the segments $\overline{pa}$ and $\overline{px}$. This angle is defined modulo 2π only, but can be made unique by fixing F_a for one x and demanding continuous changes in F_a for all neighboring x. Aleksandrov (1956), p. 47, states then:

[For] $F_a(p, C, x)$ [...] it is easy to see that it is a continuous function of all its arguments C, p, a, x simultaneously, i.e., for fixed x, a, p, C and for every $\varepsilon > 0$, there exists a $\delta > 0$ such that $F_{a'}(p', C', x')$ is defined and

$$|F_a(p, C, x) - F_{a'}(p', C', x')| < \varepsilon$$

for

$$|x' - x| < \delta, \qquad |a' - a| < \delta, \qquad |p' - p| < \delta, \qquad \rho(C', C) < \delta.$$

Here [...] $\rho(C', C) < \delta$ means that $|C'(\xi) - C(\xi)| < \delta$ for all $\xi \in [\xi_0, \xi_1]$. (Symbols adapted)

Specifically, one assumes that the curves C and C' are connected by a continuous deformation, i.e., that C and C' are *homotopic*. Clearly, $F_a(p, C, x_1) - F_a(p, C, x_0)$ is independent of a.

Replace now the real interval $[\xi_0, \xi_1]$ for the curve variable ξ by the circle S, and let $C : S \to \mathbb{R}^2$ be a closed curve. Then

$$\gamma(C, p) = \frac{1}{2\pi} \left(F_a(p, C, x_1) - F_a(p, C, x_0) \right) \in \mathbb{Z} \tag{6.11}$$

is a signed integer independent of a, x_1, x_2, termed the *index* of the point p relative to the map C, or the *winding number* of C about p.

Since F is continuous and γ an integer, the following important but rather obvious statement holds: *The index of a point p relative to a closed path C is constant under deformations of C, as long as C does not cross p*; see Aleksandrov (1956), p. 48, statement 2.24.

Note that γ is well-defined even for curves like the one shown in Fig. 6.1, for which the azimuth angle partially increases and partially decreases. The index with respect to the center point is here $\gamma = 1$, since the contribution to the angle function F_a changes sign when the curve orientation with respect to a radial ray changes. The turning points at which a ray from the central point is tangent to the curve are addressed in the exact definition of the mapping degree, see, e.g., Zeidler (1986), p. 522, or Zeidler (1976), p. 135.

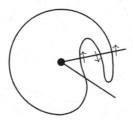

Figure 6.1   The winding number of the closed curve about the central point is 1. Multiple curve crossings along a radial ray count according to the orientation of the curve (arrows)

Instead of a smooth closed curve in the plane, we consider in the following a continuous map $A : S \to S$. Then the number $d \in \mathbb{Z}$ of coverings of the circle S by A is called the *index* or *degree* of A.

For maps $A : S^n \to S^n$ on n-spheres, $n > 1$, the algebraic details in the definition of the mapping degree (e.g., regarding the orientation of simplices in a triangulation of the sphere, and the related signs in the sum formula for the mapping degree) are somewhat subtle and we refer the reader to the literature for these. An excellent account is given in Granas & Dugundji (2003), especially their chapters 5, 8, 9 (see also Dugundji & Granas 1982).

6.3   LEMMA G1. SURJECTIVITY OF MAPS NEAR IDENTITY

Lemma G1 *Let U be a closed domain in $\mathbb{R}^n$ and $A : U \to R$ a continuous map such that*

$$|Ax - x| < \varepsilon \tag{6.12}$$

for all $x \in U$. Then (see Fig. 6.2)

$$U - \varepsilon \subset AU. \tag{6.13}$$

In other words, the map $A : U \to U - \varepsilon$ is surjective.

Proof (Arnold) The proof of this lemma in Arnold (1963a) is rather brief, and we include it here in full:

In fact, let $x_0 \in (U - \varepsilon) \backslash AU$. Then the map

$$A^* x = x_0 + \varepsilon \, |Ax - x_0|^{-1} (Ax - x_0)$$

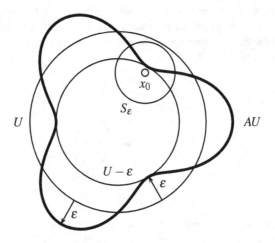

Figure 6.2 Continuous map A with $|Ax - x| < \varepsilon$ on a domain U

is continuous for $|x - x_0| < \varepsilon$. Consequently the degree $d(t)$ (Seifert & Threlfall 1934) of the map A^* of the sphere $S_{t\varepsilon} : |x - x_0| = t\varepsilon$ $(0 < t \leq 1)$ onto the sphere S_ε does not depend on t. But $d(1) = 1$, hence $x_0 \in AU$.

Arnold (1963b) is more specific toward the end of this argument:

> The degree $d(t)$ [...] does not depend on t. But $d(1) = 1$ and $d \to 0$ as $t \to 0$, i.e., $x_0 \in AU$.

The following proof is for $n = 2$ (dimension of $\mathbb{R}^n$, not number of degrees of freedom) and uses only the elementary mapping degree (index; winding number) of a curve. The case $n > 2$ is treated in the next section.

Proof $(n = 2)$ By contradiction. Assume there is an x_0 with $x_0 \in (U - \varepsilon) \backslash AU$. Shift the origin so that $x_0 = 0$ and consider the disk $B_\varepsilon(0)$ of radius ε about the origin. The map A is defined on $B_\varepsilon(0)$ including 0. Let $A^* : B_\varepsilon(0) \to S_\varepsilon$ (with 'circle' S_ε of radius ε, which is a square with side length 2ε in the maximum norm) be defined as in the quotation above,

$$A^* x = x_0 + \varepsilon \frac{Ax - x_0}{|Ax - x_0|} = \varepsilon \frac{Ax}{|Ax|}. \tag{6.14}$$

A^* is continuous on $B_\varepsilon(0)$ since $B_\varepsilon(0) \subset U$ and since A is continuous on U, and because the denominator in Eq. (6.14) cannot vanish (by assumption, there is no x

with $Ax = x_0 = 0$). New maps A_t^* with $0 < t \leq 1$ from circles $S_{t\varepsilon}$ with center 0 and radius $t\varepsilon$ to the circle S_ε are defined by

$$A_t^* : S_{t\varepsilon} \to S_\varepsilon, \qquad A_t^*(x) = A^*(x). \tag{6.15}$$

Because A is continuous, it maps closed curves to closed curves, and so A_t^* maps the circle $S_{t\varepsilon}$ to a closed curve on S_ε.

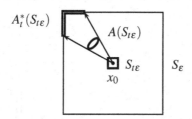

Figure 6.3 $S_{t\varepsilon}$ and its images $A(S_{t\varepsilon})$ and $A_t^*(S_{t\varepsilon})$ (bold lines) for small t

First, let $t \to 0$. The image $A(S_{t\varepsilon})$ of the infinitesimal circle $S_{t\varepsilon}$ with center x_0 is a closed loop in an infinitesimal neighborhood of Ax_0, see Fig. 6.3. Projecting this loop with straight rays emanating from x_0 to S_ε gives an infinitesimal arc $A_t^*(S_{t\varepsilon})$ (since x_0 and Ax_0 have a fixed, finite distance) on S_ε. Each point of this arc has two, four, etc., preimages on $S_{t\varepsilon}$ (with a finite number of exceptional points with an odd number of preimages); outside the arc, there are no preimages. According to the definition above, the degree of A_t^* is therefore zero (the images come in pairs with opposite curve orientation).

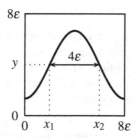

Figure 6.4 Map A_1^* with an assumed degree 0 from S_ε (abscissa) to a true subset of S_ε (ordinate). x_1 and x_2 are antipodal points

Note that for $|x_0 - Ax_0| = r$ there is always a $t \ll 1$ so that $|Ax - Ax_0| < r$ for all $x \in S_{t\varepsilon}(x_0)$, as is the case in Fig. 6.3. This implies partial covering of S_ε by $A_t^*(S_{t\varepsilon})$ and thus degree zero of A_t^*.

Second, let $t = 1$. We show that the degree of A_1^* is now different from zero. Assume instead that it is still zero, i.e., that $A_1^*(S_\varepsilon)$ covers only parts of S_ε, as for $A_t^*(S_{t\varepsilon})$ at $t \ll 1$. The image $A_1^*(S_\varepsilon)$ is then again an arc on S_ε, and each of its points, with a finite number of exceptions, has two, four, etc., preimages on S_ε.

Let l with $0 \le l \le 8\varepsilon$ be the arc length along S_ε (square with side length 2ε). Figure 6.4 shows the obvious fact that there exists an image point y of A_1^* with preimages x_1 and x_2 that are *antipodal* points, i.e., they lie on a straight line through x_0. For the following argument it is sufficient that x_1 and x_2 lie on opposite sides of the square S_ε, as shown in Fig. 6.5.

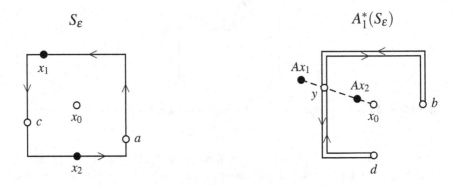

Figure 6.5 S_ε (left panel) and its image $A_1^*(S_\varepsilon)$, assumed to be a double arc with $b = A_1^*(a)$ and $d = A_1^*(c)$ (right panel). The two figures are thought to lie atop of each other. Except for b and d, each point of the arc has two preimages; specifically $A_1^*(x_1) = y = A_1^*(x_2)$ with x_1 and x_2 on opposite sides of S_ε (or even being antipodal points). Then $|Ax_2 - x_2| \ge \varepsilon$ in the max norm, violating $|Ax - x| < \varepsilon$

The image points Ax_1 and Ax_2 lie on the straight line from x_0 through y. In Fig. 6.5, the distance of each point on this segment to x_2 is $\ge \varepsilon$ (in the max norm and each other norm), violating the assumption $|Ax - x| < \varepsilon$. Thus the degree of A_1^* cannot be zero, contrary to the assumption. (It is easily seen that $|Ax - x| < \varepsilon$ implies degree one for $t = 1$. This is actually not needed.)

When t runs from 0_+ to 1, the image $A_t^*(S_{t\varepsilon})$ changes continuously. But the partially twofold (or fourfold, etc.) covering of S_ε at small t (degree 0) cannot be deformed continuously to the full covering at $t = 1$ (degree $\ne 0$, or actually 1), as is illustrated in Figure 6.6.

More formally, let $\hat{x} = x/(8t\varepsilon)$ with $0 \le \hat{x} \le 1$ be a normalized 'angle' variable on the circles $S_{t\varepsilon}$ and put $\hat{A}_t^*(\hat{x}) = A_t^*(x)$. Then $\hat{A}_t^*(\hat{x})$ is a homotopy of the functions $\hat{A}_{0_+}^*(\hat{x})$ and $\hat{A}_1^*(\hat{x})$. All the maps in a homotopy have the same degree (Brouwer 1911b, p. 105). We have thus arrived at a contradiction, $d(A_{0_+}^*) = 0$ and $d(A_1^*) \ne 0$, which implies that the point x_0 cannot exist. $\square$

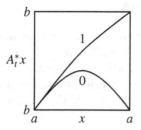

Figure 6.6 There is no continuous deformation of a function with degree 0 to a function with degree 1. The two horizontal and the two vertical sides of the square are to be identified

Remark The image $A(\partial U)$ of the boundary of U lies clearly outside the domain $U - \varepsilon$. Thus, to prove Lemma G1 one must show that AU has no 'holes' inside $U - \varepsilon$. Such holes are not forbidden by continuity: while the image of a connected domain (no disjoint open subdomains) by a continuous function is again connected, the image of a simply connected domain by a continuous function is not necessarily simply connected. (For *retracting* maps, by contrast, the image of a simply connected domain is simply connected.) Figure 6.7 demonstrates this for the map $z \mapsto e^z$ in the complex plane, $z = x + iy$, with $x \in [-1, 1]$ and $y \in [-\pi, \pi]$. The distance in the max norm between the point d and its image d' in Fig. 6.7 is $1 + e$ (in the x-direction), thus $\varepsilon = 1 + e$ in the statement of the lemma. Removing a layer of width $1 + e$ from the boundary of the rectangle $acdb$ leaves the empty set, which is in agreement with Lemma G1, $\emptyset \subset AU$, thus the central hole in the image poses no problem.

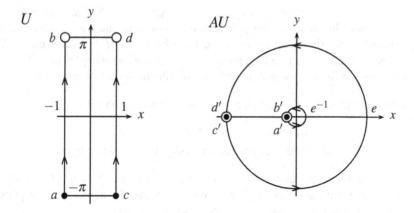

Figure 6.7 The continuous map $z \mapsto e^z$ $(z = x + iy)$ from a simply to a doubly connected domain

6.4   BROUWER'S LEMMA

In this section, Lemma G1 is proved for a map near identity in $\mathbb{R}^n$ for $n > 2$. The proof is taken from Brouwer (1911a). The following remark by Hans Freudenthal, quoted by Granas & Dugundji (2003), p. 277, stresses the importance of Brouwer's proof (this quote, almost verbatim, can also be found in the short biography of Brouwer written by H. Freudenthal and A. Heyting for volume two of Brouwer's *Collected Works*, 1976):

> Seen in historical perspective, Brouwer's performance at that time [...] looks like witchcraft. As his magic wand he used the seemingly simple theorem: *If a continuous map f of the n-dimensional cube $K^n = [-1, 1]^n$ into $\mathbb{R}^n$ displaces every point less than 1/2, then the image $f(K)$ contains an interior point.* He proved this theorem by simplicial approximation of the given mapping.

Brouwer's theorem ('Hilfssatz') becomes, after minor adaption, Arnold's Lemma G1 from the last section. Brouwer's proof is direct, whereas Arnold's proof of his lemma is by contradiction (only). Furthermore, Brouwer does not need to introduce the auxiliary maps A_t^*. Before we give Brouwer's proof, some facts are summarized about simplices, simplicial approximations and simplex orientations, following Franz (1974), Brouwer (1911b) and Hilton & Wylie (1967).

Definition A 1-simplex is an interval in $\mathbb{R}$, a 2-simplex a triangle in $\mathbb{R}^2$, a 3-simplex a tetrahedron in $\mathbb{R}^3$. A general *n-simplex* is an n-dimensional polyhedron determined by its $n + 1$ corner points. No corner shall lie in the linear subspace ('hyperplane') defined by the other corners.

An n-simplex with corners p_0 to p_n is written $[p_0 p_1 \ldots p_n]$. An *open simplex* is the set of all points that lie inside this polyhedron but not on its sides, edges and corners, i.e., in none of the (boundary) simplices defined by $m < n$ of the corner points. We universally refer to them as *sides*, and to $(n-1)$-sides of an n-simplex as its *faces*. A *closed n-simplex* is an open n-simplex together with all the points in its sides.

Definition A finite set K of simplices is a *(simplicial) complex* if

(a) for each simplex in K, each of its sides is in K,

(b) the intersection of any two open simplices in K is empty.

Two closed n-simplices can have a common face of dimension $n - 1$. Let $|K|$ be the set of all points in the complex K. Thus K consists of simplices, $|K|$ of points. For a set K whose elements σ are sets, the union is defined as

$$\bigcup K = \bigcup \{\sigma : \sigma \in K\} = \{x : \exists \sigma \; x \in \sigma \in K\}, \tag{6.16}$$

see Jech (1997), p. 6 (one inner pair of set brackets $\{\ldots\}$ is removed in the set union). Thus $|K| = \bigcup K$ (Dugundji & Granas 1982, p. 111). Note that $|K|$ is left unchanged if K is refined, i.e., its simplices are subdivided into smaller ones. For each point $x \in |K|$ the unique simplex $\sigma \in K$ with $x \in \sigma$ is called the *carrier* simplex of p. The set $|K|$ is called a *polyhedron*, and K is a *triangulation* of $|K|$. More generally, one defines:

Definition A *triangulation* or *simplicial approximation* of a topological object S (e.g., a sphere, torus, sphere with handles, Moebius strip, Klein bottle) is a simplicial complex K together with a *homeomorphism* $h : |K| \to S$ (one-to-one map h; both h and h^{-1} are continuous).

A topological object and its simplicial approximation are topologically equivalent. All differentiable manifolds can be triangulated. For dimensions $n \geq 4$, however, there exist non-differentiable topological manifolds that cannot be triangulated.

Definition A simplicial complex K is a triangulation of a *closed topological n-manifold* if (see Franz 1974, p. 38):

(a) every simplex of K is side of an n-simplex of K,

(b) every $(n-1)$-simplex of K is face of exactly two n-simplices of K,

(c) any two n-simplices of K can be connected by an alternating sequence of n- and $(n-1)$-simplices.

Here (a) ensures that a manifold has everywhere dimension n and that there are no 'collapsed' regions of lower dimensions; (b) means that a manifold has no bifurcations (as in the letter Y); and (c) ensures that a manifold is connected. If in (b) one allows each $(n-1)$-simplex to be face of one or two n-simplices, one speaks of a pseudo-manifold with boundary. The boundary is given by those $(n-1)$-simplices that are face of only one n-simplex.

Definition A *simplicial map* $\varphi : K \to L$ maps simplex corners in K to simplex corners in L so that (a) the image of every simplex in K is a single simplex in L (with possibly fewer corners). Furthermore, (b) the interior points of simplices are mapped linearly to interior points of simplices, $\varphi(\lambda_0 p_0 + \cdots + \lambda_n p_n) = \lambda_0 \varphi(p_0) + \cdots + \lambda_n \varphi(p_n)$, where p_i refers to the Cartesian coordinate tuple representing a corner in $\mathbb{R}^n$ and $\lambda_0 + \cdots + \lambda_n = 1$ and $\lambda_i > 0$ for $i = 0$ to n.

Simplicial maps are continuous and used as linear approximations to continuous maps.

Definition The simplicial map $\varphi : K \to L$ is a *simplicial approximation* to the continuous map $f : |K| \to |L|$ if for each $x \in K$ the closure of the simplex that contains $f(p)$ contains $\varphi(p)$.

A simple argument shows that $f : |K| \to |L|$ and its simplicial approximation φ are homotopic (see Hilton & Wylie 1967, pp. 36-37; for the definition of homotopy see Section 6.2). Indeed, since for each point x, the images $f(x)$ and $\varphi(x)$ lie in the closure of the same simplex, the points $f(x)$ and $\varphi(x)$ can be connected by a straight line within this closed simplex, with all points on the straight line belonging to the simplex (there are no 'holes'). Doing this for all x defines a homotopy H (H is continuous and does not leave the image space $|L|$ of f and φ). Similarly, two different simplicial approximations to a map f are homotopic; and the simplicial approximation to a simplicial map φ is φ itself.

Any n-simplex σ of a complex can be subdivided into smaller n-simplices with a new corner at the ('barycentric') midpoint of σ (for details of the combinatorics, see Hilton & Wylie 1967, pp. 22-23). Iterating this *simplex refinement* r times, one obtains a complex $K^{(r)}$.

Simplicial approximation theorem   *Let K and L be complexes and $f : |K| \to |L|$ be a continuous map. Then there is an integer $r \geq 0$ and a simplicial approximation $\varphi : K^{(n)} \to L$ to f.*

The first *proof* was given by Brouwer in 1910. For modern proofs, see Schubert (1964), pp. 173-174; Franz (1974), pp. 25-26; Hilton & Wylie (1967), p. 37. Note that L needs not be refined since it can be chosen arbitrarily fine from the start.

If a simplicial map φ maps n-simplices to n-simplices, and if two n-simplices σ_1 and σ_2 have a common face, then $\varphi(\sigma_1)$ and $\varphi(\sigma_2)$ are the same simplex or have a common face. For the n image points of the common face are identical by uniqueness of φ, and the two different corners are mapped to either one or two new corners.

We next define the *orientation* of a simplex, which is central to the mapping degree and thus to Brouwer's proof. Let $[p_0 \ldots p_n]$ be an n-simplex. The corners are considered now as an ordered sequence.

Definition   All possible $(n + 1)!$ linear orderings of the corners p_0 to p_n of an n-simplex fall into two possible classes, the even permutations of the ordered sequence $[p_0 \ldots p_n]$, called *positive orientation* $(+1)$ of the simplex, and the odd permutations, called *negative orientation* (-1). An oriented simplex is written $\langle p_0 \ldots p_n \rangle$, and $\langle p_0 p_1 p_2 \rangle = -\langle p_1 p_0 p_2 \rangle$, etc.

For a 1-simplex, the orientation is a direction, for a 2-simplex it refers to clockwise or counter-clockwise rotation, for a 3-simplex to a left- or right-handed screw (as in the right-hand rule for magnetic fields).

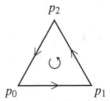

Figure 6.8   The orientation of a triangle induces orientations of its sides

The orientation of an n-simplex *induces* an orientation of its $(n-1)$-dimensional faces by the formula

$$\langle p_0 \ldots \not{p}_i \ldots, p_n \rangle = (-1)^i \langle p_0 \ldots p_i \ldots p_n \rangle, \tag{6.17}$$

where $\not{p}_i$ means that the corner p_i does not belong to the face (note that the first position has index zero). Thus the oriented triangle $\langle p_0 p_1 p_2 \rangle$ in Fig. 6.8 induces the three oriented faces $\langle p_0 p_1 \rangle$, $\langle p_1 p_2 \rangle$ and $\langle p_2 p_0 \rangle$. One has

$$\langle p_3 p_4 \rangle = \langle \not{p}_2 p_3 p_4 \rangle \quad \text{and} \quad \langle p_2 p_3 p_4 \rangle = \langle \not{p}_1 p_2 p_3 p_4 \rangle, \tag{6.18}$$

thus

$$\langle p_3 p_4 \rangle = \langle \not{p}_1 \not{p}_2 p_3 p_4 \rangle. \tag{6.19}$$

But also

$$\langle p_3 p_4 \rangle = \langle \not{p}_1 p_3 p_4 \rangle \quad \text{and} \quad \langle p_1 p_3 p_4 \rangle = -\langle p_1 \not{p}_2 p_3 p_4 \rangle, \tag{6.20}$$

thus

$$\langle p_3 p_4 \rangle = -\langle \not{p}_1 \not{p}_2 p_3 p_4 \rangle, \tag{6.21}$$

and Eqs. (6.19) and (6.21) are in contradiction since the orientation is either $+1$ or -1 (but not 0). Applying Eq. (6.17) twice or more often does therefore not result in an oriented side: an n-simplex does *not* induce an orientation of $(n-m)$-simplices with $m \geq 2$.

Definition   (Brouwer 1911b, p. 101) Let

$$\sigma_1 = \langle p_0 \ldots p_{i-1} p_i p_{i+1} \ldots p_n \rangle, \qquad \sigma_2 = \langle p_0 \ldots p_{i-1} q_i p_{i+1} \ldots p_n \rangle \tag{6.22}$$

be two oriented n-simplices with common face $[p_0 \ldots p_{i-1} p_{i+1} \ldots p_n]$ and different corners p_i and q_i. Then σ_1 and σ_2 have *opposite* orientations.

Remark 1   The crux in the definition is that the different corners p_i and q_i appear at the same position in the ordered $(n+1)$-tuples of corners. Figure 6.9 illustrates this definition for 2- and 3-simplices. Here the 2-simplices $\tau_1 = \langle p_0 p_1 p_2 \rangle$ and $\tau_2 = \langle p_3 p_2 p_1 \rangle = -\langle p_3 p_1 p_2 \rangle$ have the same orientation, as do the 3-simplices $\sigma_1 = \langle p_0 p_1 p_2 p_3 \rangle$ and $\sigma_2 = \langle p_4 p_2 p_1 p_3 \rangle = -\langle p_4 p_1 p_2 p_3 \rangle$.

Remark 2   The above definition is used in the proof of Gauss' theorem, where a given domain is subdivided into infinitesimal cells and $d\vec{a}$ switches sign when considered as surface area element of one or the other of two adjacent volume cells. The same holds for line elements $d\vec{l}$ along area cell boundaries in Stokes' theorem.

Definition   A closed topological n-manifold (a compact topological manifold without boundary) is *orientable* if all the n-simplices of a simplicial approximation can be oriented in such a way that any two n-simplices σ_1 and σ_2 induce *opposite* orientations of a common face.

A sphere and a torus are orientable, the Moebius strip and the Klein bottle are not.

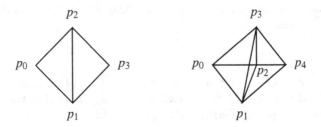

Figure 6.9   Adjacent 2- and 3-simplices

In a simplicial map $\varphi : K \to L$, a given image simplex in L may have $1, 2, 3 \ldots$ different preimage simplices $\sigma_i \in K$. A simplicial map is linear and maps corners to corners, but there is no restriction on the order of the image corners, thus simplicial maps in general do not preserve orientation. A given simplex may occur with both positive and negative orientations in the image of a simplicial map φ. The respective multiplicities are named q_+ and q_-. The next lemma is central to Brouwer's proof.

Lemma *If two n-simplices σ_1 and σ_2 with the same orientation have a common face and the simplicial map φ maps n-simplices to n-simplices, then $\varphi(\sigma_1)$ and $\varphi(\sigma_2)$ are either two different simplices with the same orientation and a common face, or twice the same n-simplex with opposite orientations.*

Remark The orientation of the two image simplices may be opposite to that of σ_1 and σ_2.

Proof After renaming corners, p_0 and p_1 can be assumed as different corners, and p_2 to p_{n+1} as common corners of σ_1 and σ_2. In agreement with Eq. (6.22), let $\sigma_1 = \langle p_0 p_2 p_3 \ldots p_{n+1} \rangle$ and $\sigma_2 = -\langle p_1 p_2 p_3 \ldots p_{n+1} \rangle$ (same orientation). Then for $f(p_0) = P_0 = f(p_1)$ one has

$$\langle p_0 p_2 p_3 \ldots p_{n+1} \rangle \overset{f}{\mapsto} \langle P_0 P_2 P_3 \ldots P_{n+1} \rangle \tag{6.23}$$

and

$$-\langle p_1 p_2 p_3 \ldots p_{n+1} \rangle \overset{f}{\mapsto} -\langle P_0 P_2 P_3 \ldots P_{n+1} \rangle, \tag{6.24}$$

thus the single image simplex $[P_0 P_2 P_3 \ldots P_{n+1}]$ appears once with positive and once with negative orientation. In the second case, $f(p_0) = P_0 \neq P_1 = f(p_1)$, Eq. (6.23) remains unchanged and Eq. (6.24) becomes

$$-\langle p_1 p_2 p_3 \ldots p_{n+1} \rangle \overset{f}{\mapsto} -\langle P_1 P_2 P_3 \ldots P_{n+1} \rangle. \tag{6.25}$$

According to Eq. (6.22), the image simplices in Eqs. (6.23) and (6.25) have the same orientation (different corners at the same position in $\langle \ldots \rangle$). $\qquad \square$

One may think of the case that two preimage simplices are mapped to one image simplex with different orientations as folding the line, surface, volume, etc., formed by the image simplices back upon itself.

This finishes the summary on simplices and their orientations. The following lemma by Brouwer (1911a) is identical to Arnold's Lemma G1 above, except that the domain U is arbitrary there, but is assumed to be a cube here. Quite surprisingly, Arnold (1963a,b) does not use or refer to Brouwer's lemma. Brouwer's original proof is also discussed in Dieudonné (1989), pp. 169-173.

Lemma (Brouwer 1911a) *Let U be a cube with dimension n and side length a, and $f : U \rightarrow \mathbb{R}^n$ be a continuous map with $\|f - E\| \leq \varepsilon < a/2$. Then $U - \varepsilon \subset f(U)$.*

Remark The surprising fact is that $f(U)$ cannot have a central hole.

Proof (i) Let the complexes K and L be triangulations of U and $f(U)$, respectively, and $\varphi : K \rightarrow L$ be a simplicial approximation to f. If the image of an n-simplex is an m-simplex, $m < n$ (simplex collapse), move corners of K slightly to obtain image dimension n; this will not change the conclusions.[1] All n-simplices of K shall be oriented in the positive sense ($+1$; the cube is orientable), the image n-simplices of φ have orientation $+1$ and -1. Each *boundary* face in ∂U (perpendicular hyperplanes) belongs to one n-simplex, each *interior* face of the cube to two n-simplices.

(ii) Let V be the set of points of the small cube $U - \varepsilon \neq \emptyset$ concentric with U that do not lie in the φ-image of an $(n - m)$-simplex of K with $m \geq 2$ (they have no orientation, thus the following argument does not apply to them). Any two points of V can be connected by straight line segments l that lie fully in V (cross no $(n - m)$-simplex). Still, this line may cross faces. Since $\|f - E\| \leq \varepsilon$, the images of boundary faces of K in ∂U cannot lie in $U - \varepsilon$ and thus not in V. Let $W \subset V$ be those points that do not lie in the φ-image of an interior face of K.

(iii) Let $P_1, P_2 \in W \subset U - \varepsilon$ be two arbitrary points connected by straight segments $l \subset V$, and let $P \in W$ be an arbitrary point on l. If P lies in an image simplex $\sigma \in L$ (carrier simplex) under the map φ, then σ shall have multiplicity $M = q_+ + q_-$, with q_+ positive and q_- negative orientations. When P moves along l in the inner cube $U - \varepsilon$, the numbers q_+ and q_- can only change if P crosses the φ-image of an interior face of K.

(iv) When P crosses such a face, three cases are possible: (a) the two φ-image n-simplices σ_1, σ_2 with the common face are adjacent simplices with the same orientation, thus neither q_+ nor q_- changes when P passes from one to the other. Alternatively, $\sigma_1 = -\sigma_2$ is the same simplex with opposite orientations. Then P moves (b) either into a new pair of φ-image simplices, with multiplicity $M \rightarrow M + 2$ and both q_+ and q_- increase by 1, or (c) leaves such a pair, $M \rightarrow M - 2$ and both q_+ and q_- decrease by 1. In all three cases (a) to (c), the signed integer $d = q_+ - q_-$, called the

---

[1] See Brouwer (1911a) for the simple argument demonstrating this.

mapping degree, remains constant, and is thus constant on the path l of P from P_1 to P_2 and has the same value in all points of W.

(v) Finally, one has to show that $d = 1$. The argument is novel, and we quote Brouwer directly:

> To find the value of this constant d, we denote the content [volume or measure] of $U - \varepsilon$ with I and notice that the total content of those parts of the image simplices that are contained in $U - \varepsilon$ is $d \cdot I$. If we then continuously decrease the displacements of the base points [corners] that underlie the simplicial map φ until they all reach the value zero,[2] and accordingly let the map φ turn into the identity map, then the total content cannot suffer any jumps, and thus the whole number d cannot change. From this we conclude that the number d has the same value for φ as for the identity map, i.e., the value 1. (Brouwer 1911a, p. 163; symbols adapted)[3]

Therefore d of φ equals d of E, which is 1, and $W \subset U - \varepsilon$ lies in the image of φ. Furthermore, if an open image simplex lies in the closed domain $U - \varepsilon$ (see text before Eq. 2.20), the same is true for its closure. Finally, since by simplex refinement f can be approximated arbitrarily close by φ, we conclude that $U - \varepsilon \subset f(U)$.   □

As Dieudonné (1989), p. 171, notes, '[t]he introduction of the volume of I is not at all artificial.' It also appears in Milnor's 'strange but quite elementary proof' (Milnor 1978, p. 521) of Brouwer's fixed point theorem and the so-called 'hairy ball theorem.'

We close this section with a discussion of Brouwer's lemma. The latter states that a continuous map $f : K \to \mathbb{R}^n$, with K the n-dimensional cube with side length 1 centered at the origin, fully contains in its range the n-dimensional cube L with side length $1 - 2a$ centered at the origin, if all the displacements due to f are smaller than $a \le 1/2$. The two crucial properties of f are continuity and proximity to E. There are four possible ways that f may fail the lemma: (i) the f-displacements open holes inside L; (ii) the image of f 'wraps' around a hole in L, as in the complex exponential map; (iii) the image of f has dimension $< n$ in parts of L; (iv) the image of f has topological 'anomalies' like gaps.

The latter two items were a serious challenge at Brouwer's time, after Cantor (1877) had found a bijection between $\mathbb{R}$ and $\mathbb{R}^2$ (see Gouvêa 2011 for a historical account), and Peano (1890) had defined a space-filling curve. The concept of

---

[2] 'f is homotopic to the identity in K.' (Dieudonné 1989, p. 180)

[3] 'Um den Wert dieser Konstante d zu ermitteln, bezeichnen wir den Inhalt von $U - \varepsilon$ mit I und bemerken, dass der Gesamtinhalt derjenigen Teile der Bildsimplexe, welche in $U - \varepsilon$ enthalten sind, $d \cdot I$ beträgt. Wenn wir dann die der simplizialen Abbildung φ zugrunde liegenden Verrückungen der Grundpunkte stetig verkleinern, bis sie alle den Wert Null erreichen, und die Abbildung φ entsprechend stetig in die identische Abbildung übergehen lassen, so kann dabei der genannte Gesamtinhalt keine Sprünge erleiden, und somit die ganze Zahl d sich nicht ändern. Hieraus folgern wir, dass die Zahl d für φ denselben Wert, wie für die identische Abbildung, d. h. den Wert 1, besitzt.'

a *nowhere dense* set (see Section 3.7) was well-established at Brouwer's time; he uses the term in his paper (1911a). Brouwer's way to overcome these difficulties with the continuum and its dimension is simple and direct: he triangulates domains (manifolds) using simplices of small but finite size. The problematic continuum is put inside the simplex and handled by considering linear simplicial maps as approximation to the given map. One needs to deal then only with the combinatorics of a finite number of simplex corners when discussing the properties of this map.

Brouwer's proof has two key features: first, the introduction of a mapping degree $d = q_+ - q_-$ that characterizes a function by the number of its preimages of a given orientation. Second, the observation that two adjacent n-simplices in K with the same orientation are either mapped to two adjacent n-simplices (after mapping to simplices of lower dimension is avoided), again of the same orientation; or to the same image simplex twice, however with opposite orientations. This uses continuity of f, otherwise simplicial maps cannot be defined (inside a simplex, they are linear hence continuous, thus any discontinuity must occur at a corner, mapping it non-sensically to two different corners). By moving through the complex L along straight segments and crossing $(n-1)$-dimensional faces, one either moves from n-simplex to n-simplex with identical orientations, or one loses or gains a pair of image simplices with opposite orientations; whatever is the case, d stays constant.

This gives a clear geometric picture of the possible behavior of f. One starts with the inner cube L covered once with simplices and points, i.e., $d = 1$ for each $x = E(x)$. The displacements caused by f (say, in a sequence of single corner-to-corner shifts) cannot 'tear holes' into L, since simplices can leave L only in pairs with opposite orientations, but one started with $d = 1$. Still f can move 'folds' (double layers) of simplices *into* L, which however does not change $d = 1$. f cannot 'unveil' L, since the displacements are too small to move a K-boundary into L. f cannot collapse domains to lower dimension, since n-simplices are mapped to n-simplices. Finally, f cannot cause topological anomalies like gaps, since it is approximated by a linear map inside simplices. This answers objections (i) to (iv) from above.

6.5 MODERN PROOF OF LEMMA G1

In this section we give a modern proof of Lemma G1. Consider the map A_1^* defined in Eq. (6.15),

$$A_1^* : S_\varepsilon \to S_\varepsilon, \qquad A_1^*(x) = \varepsilon \, \frac{Ax}{|Ax|}, \tag{6.26}$$

where $|x| = \varepsilon$ and $\|A_1^* - E\| < \varepsilon$. It suffices to show that the degree of A_1^* is odd (i.e., not zero), since then the degree of all A_t^* for $0 < t \le 1$ is odd because they are homotopic to A_1^* (according to Brouwer 1911b, two homotopic maps have the same degree), and thus also A^* (the 'union' of all A_t^*) has odd degree.

A_1^* maps the $(n-1)$-sphere S^{n-1}, i.e., the boundary of a ball centered at the origin in $\mathbb{R}^n$, to itself. The study of such maps is a major topic in *fixed point theory*. We assume a basis e_1 to e_n of $\mathbb{R}^n$ and, without loss of generality, radius 1 for S^{n-1}.

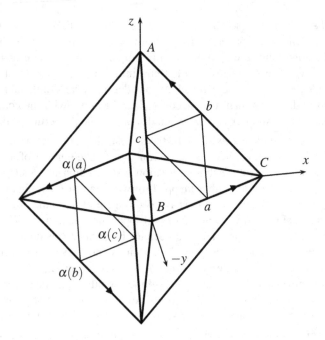

Figure 6.10 Triangulation of the sphere $S^2 \subset \mathbb{R}^3$. α is the antipodal map, and arrows indicate corner displacements by an antipodal-preserving simplicial map φ

Definition The *basic triangulation* Σ^{n-1} of the sphere S^{n-1} consists of the 2^n simplices $[\pm e_1, \pm e_2, \ldots, \pm e_n]$ of dimension $n-1$ and all their sides. Figure 6.10 shows this triangulation for the 2-sphere in $\mathbb{R}^3$ (a double pyramid), consisting of 8 triangles. As a specific simplex of the basic triangulation one chooses $[e_1, -e_2, e_3, \ldots, (-1)^{n-1} e_n]$; in the figure, this is the triangle $[ABC]$. The *antipodal* map $\alpha : S^{n-1} \to S^{n-1}$ is defined by $\alpha(x) = -x$ for arbitrary x. This can be generalized to simplices by $\alpha([e_1 e_2 \ldots]) = [\alpha(e_1)\alpha(e_2)\ldots]$. With each simplex σ, the basic triangulation also contains its antipodal simplex $\alpha(\sigma)$. A triangulation is called *symmetric* if it has this property, and if all spheres S^k with $k < n-1$ are unions of simplices of this triangulation; e.g., in Fig. 6.10, the circle S^1 is the union of the simplex $[BC]$ and the corresponding three lines in the xy-plane. A simplicial map φ is *antipodal preserving* if

$$\varphi = \alpha \circ \varphi \circ \alpha, \tag{6.27}$$

or, since $\alpha \circ \alpha = E$, equivalently $\varphi \circ \alpha = \alpha \circ \varphi$. The following lemma is fundamental.

Combinatorial lemma *Let* $\varphi : K \to \Sigma^{n-1}$ *be an antipodal preserving simplicial map from a symmetric triangulation K of S^{n-1} to the basic triangulation Σ^{n-1}. Then φ maps an* odd *number of simplices of K onto the simplex $[e_1, -e_2, \ldots, (-1)^{n-1} e_n]$ in Σ^{n-1}.*

The *proof*, about one page of combinatorics, is given in Dugundji & Granas (1982), pp. 40-41, from which all the material in the present section is taken.

Figure 6.10 gives an example for the lemma: the corners a, b, c define a subdivision of the simplex $[ABC]$ into four smaller triangles. Doing this for all 8 faces, one obtains a triangulation of S^2 with 32 triangles. The filled arrows ► in the figure show displacements by the simplicial map φ, which are $\varphi(a) = C$ and $\varphi(b) = A$ and $\varphi(c) = B$ together with $\varphi(A) = A, \varphi(B) = B, \varphi(C) = C$; correspondingly for the other $[\pm e_x, \pm e_y, \pm e_z]$. One easily sees that φ is antipodal preserving. Three of the small triangles in $[ABC]$ are mapped by φ to one-dimensional simplices $[AB], [BC], [AC]$ of Σ_2, but $\varphi([abc]) = [ABC] = [e_1, -e_2, e_3]$, as stated in the combinatorial lemma.

Note that any map f for which φ is a simplicial approximation has odd degree on S^{n-1}.

Theorem (Borsuk 1933) *If a continuous map $f : S^{n-1} \to S^{n-1}$ has $f(x) \neq f(\alpha(x))$ for all points, i.e., no pair of antipodal points has the same image point, then the degree of f is odd.*

Proof (see Dugundji & Granas 1982, p. 142) Choose a symmetric triangulation of the sphere, so that with each simplex its antipodal simplex is also contained in the triangulation. Clearly, symmetric triangulations exist for arbitrary (barycentric) refinements of simplices. From $f(\alpha(x)) \neq f(x) = \alpha(\alpha(f(x)))$ one has that $f(\alpha(x))$ and $\alpha(f(x))$ are not antipodal points. Thus the straight segment connecting them does not go through the origin. Project this segment from the origin onto S^{n-1}. Doing this for all points x defines a homotopy F between $f \circ \alpha$ and $\alpha \circ f$, given by

$$F(x,t) = \frac{(1-t)f(\alpha(x)) + t\alpha(f(x))}{|(1-t)f(\alpha(x)) + t\alpha(f(x))|}. \tag{6.28}$$

Let $\Phi(x) = F(x, \frac{1}{2})$, then $\alpha \circ f$ is homotopic to

$$\Phi(x) = \frac{f(\alpha(x)) + \alpha(f(x))}{|f(\alpha(x)) + \alpha(f(x))|}. \tag{6.29}$$

Clearly, $\alpha(\Phi(x)) = \Phi(\alpha(x))$, thus Φ is antipodal preserving. Furthermore $\Phi(x) = \alpha(\Phi(x))$, thus $\alpha \circ \Phi$ is also antipodal preserving. Since $\alpha \circ f$ is homotopic to Φ, $f = \alpha \circ \alpha \circ f$ is homotopic to $\alpha \circ \Phi$. According to the combinatorial lemma, the degree of $\alpha \circ \Phi$ is odd, thus the degree of f is odd. $\qquad\square$

Corollary (Lemma G1) *The degree of A_1^* from Eq. (6.26) is not zero.*

Proof Assume that the degree is zero. Then according to Borsuk's theorem there is a point x so that x and $\alpha(x)$ are mapped to the same point y on S_ε with radius ε. Then clearly either $|y - x| \geq \varepsilon$ or $|y - \alpha(x)| \geq \varepsilon$, violating $\|A_1^* - E\| < \varepsilon$. $\qquad\square$

6.6   LEMMA G2. FREQUENCY DIFFEOMORPHISM

The next lemma treats the inversion of the ε-displacement.

Lemma G2 *As in Lemma G1, let U be a closed domain in $\mathbb{R}^n$ and $A : U \to \mathbb{R}^n$ a continuous map with $|Ax - x| < \varepsilon$ for all $x \in U$. Furthermore, for all x, it shall hold that*

$$|dx| \neq 0 \quad \to \quad |dA| \neq 0. \tag{6.30}$$

Then A is a diffeomorphism of the domain $U - 4\varepsilon$.

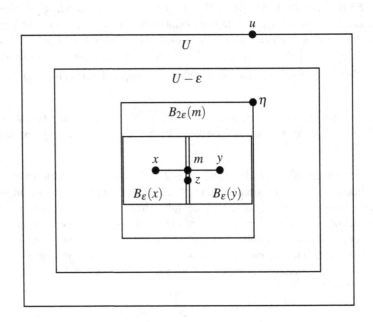

Figure 6.11 Domain $U - 4\varepsilon$

Proof (i) A is a map, thus is differentiable. To show that A is a diffeomorphism, it suffices to show that A is injective (that A is surjective was shown in Lemma G1). Let $x, y \in U - 4\varepsilon$ with $x \neq y$ and assume that $Ax = z = Ay$. Let $\overline{xy}$ be the straight line segment connecting x and y. The distance between x and y is

$$|x - y| = |x - z + z - y| = |x - Ax - (y - Ay)| \leq |x - Ax| + |y - Ay| < 2\varepsilon \tag{6.31}$$

by assumption on A. A ball $B_{2\varepsilon}(m)$ of radius 2ε about the midpoint m of $\overline{xy}$ contains all points $A\xi$ for arbitrary $\xi \in \overline{xy}$, see Fig. 6.11. The distance between x and a point η on the boundary of $B_{2\varepsilon}(m)$ is

$$|x - \eta| = |x - m + m - \eta| \leq |x - m| + |m - \eta| < \varepsilon + 2\varepsilon = 3\varepsilon. \tag{6.32}$$

This holds for the distance between any point on $\overline{xy}$ and any boundary point of $B_{2\varepsilon}(m)$. For any norm, $\big| |a| - |b| \big| \leq |a - b|$ holds. Thus, the distance of boundary points $\eta \in \partial B_{2\varepsilon}(m)$ and $u \in \partial U$ is

$$
\begin{aligned}
|u - \eta| &= |u - x - (\eta - x)| \\
&\geq |\,|u - x| - |\eta - x|\,| \\
&= |u - x| - |\eta - x| \\
&> \varepsilon,
\end{aligned}
\tag{6.33}
$$

because $|u - x| \geq 4\varepsilon$ (since $x \in U - 4\varepsilon$) and $|\eta - x| < 3\varepsilon$ from Eq. (6.32). Therefore, $\eta \in U - \varepsilon$, which is then also true for all interior points of $B_{2\varepsilon}(m)$. According to Lemma G1, all points of $B_{2\varepsilon}(m)$ are image points of $A : U \to R$, since $B_{2\varepsilon}(m) \subset U - \varepsilon \subset AU$.

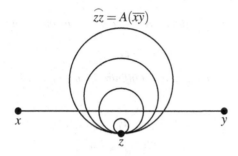

$$\widehat{zz} = A(\overline{xy})$$

Figure 6.12   Shrinking loops zz. The loop diameter converges to 0, as must $|x - y|$

(ii) The image $A(\overline{xy})$ of the straight line segment $\overline{xy}$ is a closed curve, e.g., a loop $\widehat{zz}$ that starts and ends in z and lies fully in $B_{2\varepsilon}(m)$ (Fig. 6.12). One can continuously shrink this loop $\widehat{zz}$ to the point z, staying within $B_{2\varepsilon}(m)$. Thus according to (i), all points of these smaller and smaller loops $\widehat{zz}$ are image points of A. Since $\widehat{zz}$ starts and ends in z, with $Ax = z = Ay$, one has by continuity

$$\widehat{zz} = A(\widehat{xy}), \tag{6.34}$$

where the (unknown) curves $\widehat{xy}$ connect x and y. For the shrinking loops $\widehat{zz}$, $\sup_{\zeta} |\zeta - z|$ with $\zeta \in \widehat{zz}$ converges to 0, which can be written as $|dA| \to 0$. According to Eq. (6.30), this implies $|dx| \to 0$ for $\sup_{\xi} |\xi - x|$ for $\xi \in \widehat{xy}$. Since the latter supremum is an upper bound for $|x - y|$, one has $x = y$.   □

Remark There is a serious error in the English translation of Arnold (1963a), where the condition of the above lemma is given as '$|dx| \neq 0$ whenever $|dA| \neq 0$,' or $|dA| \neq 0 \to |dx| \neq 0$, which is in the opposite direction as in Eq. (6.30), $|dx| \neq 0 \to |dA| \neq 0$. The literal translation of the Russian original is here 'for $dx \neq 0$ always $dA \neq 0$,' or $|dx| \neq 0 \to |dA| \neq 0$, as in Eq. (6.30). Furthermore, in Arnold (1963b), the English translation of this condition reads 'd$A \neq 0$ for d$x \neq 0$,' which is the literal translation of the Russian original; thus in Arnold (1963b) also $|dx| \neq 0 \to |dA| \neq 0$.[4] Condition (6.30) is evidently necessary: in order that A is one-to-one, it cannot have extrema, thus $|dx| \neq 0 \to |dA| \neq 0$.

6.7   LEMMA G3. CANONICAL TRANSFORMATION

The core of Arnold's proof of the KAM theorem is an infinite sequence of infinitesimal canonical transformations that transforms the slightly perturbed system to an integrable system. The next lemma addresses a single infinitesimal canonical transformation in this sequence. Into the proof of Lemma G3 is nested the simple proof of an auxiliary theorem on determinants.

Lemma G3 *Let G and U be domains of two n-dimensional complex spaces for variables P_i and q_i, and*

$$S(P,q) : G \times U \to \mathbb{C} \tag{6.35}$$

be an analytic function that defines an infinitesimal canonical transformation by Eq. (1.26),[5]

$$\begin{aligned} p_i &= P_i + S_{q_i}, \\ Q_i &= q_i + S_{P_i}. \end{aligned} \tag{6.36}$$

If

$$\|S\| \leq M \leq \frac{\beta^2}{16n}, \tag{6.37}$$

then the map

$$B : G - 2\beta \times U - 3\beta \to \mathbb{C}^{2n}, \qquad (P,Q) \mapsto (p,q) \tag{6.38}$$

with (p,q) and (P,Q) according to Eq. (6.36) has the following properties:

(I) *B is a diffeomorphism,* (6.39)

(II) *$\|B - E\| \leq \dfrac{M}{\beta}$,* (6.40)

(III) *$\|dB\| < 2\,|dX|$ for $X = (P,Q)$.* (6.41)

If the domain of Q is further restricted, from $U - 3\beta$ to $U - 3\beta - \delta$, one has

(IV) *$|P - p| \leq \dfrac{M}{\delta}$ for $Q \in U - 3\beta - \delta$.* (6.42)

---

[4] As expressed by Arnold in Khesin & Tabachnikov (2014), p. 28, '[...] if you are trying to find a Russian paper and if in the Russian paper it was written that *A implies B*, then in the translation [...] you will usually find that *A is implied by B*.'

[5] We follow Arnold and use the symbol S instead of s in the following.

Proof (i) With the restrictions $P \in G - \beta$ and $q \in U - \beta$ (these are $2n$ restrictions for coordinates $P_1, \ldots, P_n, q_1, \ldots, q_n$), one has for $\|S\| \le M$ from the Cauchy estimate in Eq. (5.33) that the $2n$ derivatives S_{P_i} and S_{q_i} are bound by

$$|S_{P_i}(P,q)| \le \frac{M}{\beta}, \qquad\qquad |S_{q_i}(P,q)| \le \frac{M}{\beta}. \qquad (6.43)$$

Furthermore, for any X,

$$|(B - E)X| = |(p,q) - (P,Q)| = |(S_q(P,q), -S_P(P,q))| \le \frac{M}{\beta} \le \frac{\beta}{16n} < \frac{\beta}{5}, \quad (6.44)$$

which is statement (II) of the lemma (on a larger domain). The estimate $\beta/5$ will be used below, to obtain a domain $U - \beta - 4 \times 0.2\beta - 0.2\beta = U - 2\beta$.

(ii) We construct a diffeomorphism $\tilde{B} : q \mapsto Q$. For each $P \in G - \beta$, treated as a parameter, let

$$\tilde{B} : q \mapsto Q = q + S_P. \qquad (6.45)$$

Then, for given P,

$$\|(\tilde{B} - E)q\| = |S_P(P;q)| \le \frac{M}{\beta} < \frac{\beta}{5}. \qquad (6.46)$$

The Cauchy estimate in Eq. (5.33) gives for each of the second derivatives $S_{P_i q_j}$ for $P \in G - \beta$ and $q \in U - \beta$,

$$|S_{P_i q_j}(P,q)| \le \frac{2M}{\beta^2} \le \frac{1}{8n}. \qquad (6.47)$$

One can furthermore assume that $\max_{i,j} |S_{P_i q_j}| > 0$. Else Eq. (6.36) would give $\partial Q_i / \partial q_i = 1 + S_{P_i q_i} = 1$, or $q_i = Q_i$, thus there would be no infinitesimal transformation (except for an irrelevant offset). From Eq. (6.45), one has the vector equation

$$d\tilde{B}(q) = \left(\sum_{j=1}^{n} \tilde{B}_{1,q_j} \, dq_j, \ldots, \sum_{j=1}^{n} \tilde{B}_{n,q_j} \, dq_j \right)$$

$$= \left(dq_1 + \sum_{j=1}^{n} S_{P_1 q_j} dq_j, \ldots, dq_n + \sum_{j=1}^{n} S_{P_n q_j} dq_j \right), \qquad (6.48)$$

or, in matrix notation (using column vectors now),

$$\begin{pmatrix} d\tilde{B}_1 \\ \vdots \\ d\tilde{B}_n \end{pmatrix} = \begin{pmatrix} 1 + S_{P_1 q_1} & S_{P_1 q_2} & \cdots & S_{P_1 q_n} \\ S_{P_2 q_1} & 1 + S_{P_2 q_2} & \cdots & S_{P_2 q_n} \\ \vdots & \vdots & \ddots & \vdots \\ S_{P_n q_1} & S_{P_n q_2} & \cdots & 1 + S_{P_n q_n} \end{pmatrix} \cdot \begin{pmatrix} dq_1 \\ \vdots \\ dq_n \end{pmatrix}. \qquad (6.49)$$

All diagonal elements have a lower bound $|1 + S_{P_i q_i}| \ge 1 - \varepsilon$ with $\varepsilon = 1/(8n)$, and all off-diagonal elements $i \ne j$ have an upper bound $|S_{P_i q_j}| \le \varepsilon$. Intuitively, the determinant of such a matrix should not vanish for sufficiently small ε. The next theorem shows that, for an $(n \times n)$-matrix, this is indeed correct if $\varepsilon < 1/n$.

Theorem *If an $(n \times n)$-matrix (a_{ij}) with complex numbers a_{ij} obeys*

$$|a_{ii}| > \sum_{j \neq i} |a_{ij}|, \qquad\qquad i = 1, \ldots, n, \qquad\qquad (6.50)$$

then its determinant $|a_{ij}|$ does not vanish.

Proof (Taussky[6] 1949) Assume that $|a_{ij}| = 0$. Then the system of equations

$$a_{11} x_1 + \cdots + a_{1n} x_n = 0,$$

$$\vdots$$

$$a_{n1} x_1 + \cdots + a_{nn} x_n = 0 \qquad\qquad (6.51)$$

has a non-trivial solution with $\max(|x_1|, \ldots, |x_n|) = |x| > 0$. Assume that x_i is the component or one of the components for which $|x_i| = |x|$. The i-th equation in System (6.51) is

$$-a_{ii} x_i = \sum_{j \neq i} a_{ij} x_j. \qquad\qquad (6.52)$$

Taking the abs on both sides gives

$$|a_{ii}| \, |x| \leq \sum_{j \neq i} |a_{ij}| \, |x_j| \leq |x| \sum_{j \neq i} |a_{ij}|, \qquad\qquad (6.53)$$

and therefore

$$|a_{ii}| \leq \sum_{j \neq i} |a_{ij}|, \qquad\qquad (6.54)$$

in contradiction to the assumption of the theorem, Eq. (6.50). Thus $|a_{ij}| \neq 0$. Since any of the components x_1 to x_n can give the maximum $|x|$, one must assume the condition in Eq. (6.50) for each of the n rows of the matrix, in order to deduce this contradiction in each possible case. $\qquad\qquad\square$

For the matrix in Eq. (6.49) with $|S_{P_i q_j}| \leq \varepsilon$, Condition (6.50) becomes $|1 - \varepsilon| > (n-1)\,\varepsilon$, or $\varepsilon < 1/n$, as stated above. This theorem shows that *diagonally dominant matrices* as defined by Eq. (6.50) have non-vanishing determinant. A finite lower bound for the abs of the determinant was given by Ostrowski (1938), and recently improved by Brent et al. (2014).

Since the determinant of the matrix in Eq. (6.49) does therefore not vanish, the corresponding homogeneous system $d\tilde{B}_i = 0 = \sum_j (\delta_{ij} + S_{P_i q_j})\, dq_j$ has *only* the trivial solution $|dq| = 0$. Thus we conclude that $|dq| > 0$ implies $|d\tilde{B}| > 0$.[7]

---

[6] 'Proofs [of this theorem] are being published again and again; [...] some of the proofs that have been given are very complicated' (Taussky 1949, p. 672). This is Olga Taussky-Todd, friend of Kurt Gödel in Vienna, member of the Vienna Circle, co-editor of Hilbert's *Collected Works*, co-worker of Emmy Noether and later professor at Caltech.

[7] In Arnold (1963a,b), $|d\tilde{B}| > 0$ is stated as a direct consequence of Eqs. (6.43) and (6.47) without further proof.

The map $\tilde{B}$ thus fulfills all the assumptions of the last two Lemmas G1 and G2, i.e., continuity, smallness (with $\varepsilon = \beta/5$ from Eq. 6.46) and $|dq| > 0 \rightarrow |d\tilde{B}| > 0$. According to Lemma G2 then, $\tilde{B}$ is a diffeomorphism of the domain $q \in (U - \beta) - \frac{4}{5}\beta = U - 1.8\beta$. By Lemma G1, the image of the diffeomorphism $\tilde{B}$ of the domain $q \in U - 1.8\beta$ contains the domain $Q \in (U - 1.8\beta) - 0.2\beta = U - 2\beta$. Hence for each $P \in G - \beta$ (treated as a parameter) and image point $Q \in U - 2\beta$ one has an inverse map

$$Q \mapsto q = \tilde{B}^{-1}(Q). \tag{6.55}$$

This three-step construction of the domain $U - 2\beta$, in order to ensure that (a) the Cauchy estimate holds, (b) $\tilde{B}$ is a diffeomorphism, and (c) $\tilde{B}^{-1}$ acts on the image of $\tilde{B}$, is shown in Fig. 6.13.

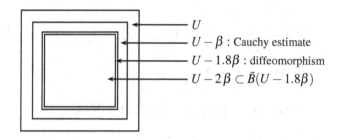

Figure 6.13 Construction of the domain $U - 2\beta$

(iii) We have thus obtained a *diffeomorphism* (statement (I) of the lemma)

$$B: G - \beta \times U - 2\beta \rightarrow \mathbb{C}^{2n} \tag{6.56}$$

with

$$B(P,Q) = (p,q) = (P + S_q(P,q), \tilde{B}^{-1}(Q)) = (P + \tilde{S}_Q(P,Q), \tilde{B}^{-1}(Q)). \tag{6.57}$$

Here the variable q in $S_q(P,q)$ is replaced, after performing the q-differentiation, by $\tilde{B}^{-1}(Q)$, to make the image (p,q) of B depend on the free variables (P,Q). This defines the new function $\tilde{S}_Q(P,Q) = S_q(P,q)$. The expressions $\tilde{S}, \tilde{S}_P, \tilde{S}_{PP}, \tilde{S}_{PQ}$ and $\tilde{S}_{QQ}$ are defined correspondingly, especially

$$q_i = Q_i - S_{P_i}(P,q) = Q_i - \tilde{S}_{P_i}(P,Q). \tag{6.58}$$

We restrict now the domain of B from $P \in G - \beta$ and $Q \in U - 2\beta$ to $P \in G - 2\beta$ and $Q \in U - 3\beta$. Since S and $\tilde{S}$ are analytic and $|S| \leq M, |\tilde{S}| \leq M$, the Cauchy estimate (5.33) applies, giving

$$|\tilde{S}_{P_iP_j}|, |\tilde{S}_{P_iQ_j}|, |\tilde{S}_{Q_iQ_j}| \leq \frac{2M}{\beta^2} \leq \frac{1}{8n}. \tag{6.59}$$

Therefore, for arbitrary $X = (P, Q)$,

$$|dB(X) - dX| =$$

$$\overset{(a)}{=} \left| \left(\sum_{j=1}^{n} \frac{\partial p_i}{\partial P_j} \, dP_j + \sum_{j=1}^{n} \frac{\partial p_i}{\partial Q_j} \, dQ_j, \ \sum_j \frac{\partial q_i}{\partial P_j} \, dP_j + \sum_j \frac{\partial q_i}{\partial Q_j} \, dQ_j \right) - (dP_i, dQ_i) \right|$$

$$\overset{(b)}{=} \left| \left(\sum_j (\delta_{ij} + \tilde{S}_{Q_iP_j}) \, dP_j - dP_i + \sum_j \tilde{S}_{Q_iQ_j} \, dQ_j, \right. \right.$$

$$\left. \left. - \sum_j \tilde{S}_{P_iP_j} \, dP_j + \sum_j (\delta_{ij} - \tilde{S}_{P_iQ_j}) \, dQ_j - dQ_i \right) \right|$$

$$= \left| \left(\sum_j \tilde{S}_{Q_iP_j} \, dP_j + \sum_j \tilde{S}_{Q_iQ_j} \, dQ_j, \ -\sum_j \tilde{S}_{P_iP_j} \, dP_j - \sum_j \tilde{S}_{P_iQ_j} \, dQ_j \right) \right|$$

$$\overset{(c)}{=} \max \left(\left| \sum_j \tilde{S}_{Q_iP_j} \, dP_j + \sum_j \tilde{S}_{Q_iQ_j} \, dQ_j \right|, \ \left| \sum_j \tilde{S}_{P_iP_j} \, dP_j + \sum_j \tilde{S}_{P_iQ_j} \, dQ_j \right| \right)$$

$$\overset{(d)}{=} \left| \sum_j \tilde{S}_{Q_1P_j} \, dP_j + \sum_j \tilde{S}_{Q_1Q_j} \, dQ_j \right|$$

$$\leq \max_j (|dP_j|) \sum_j |\tilde{S}_{Q_1P_j}| + \max_j (|dQ_j|) \sum_j |\tilde{S}_{Q_1Q_j}|$$

$$= |dP| \sum_{j=1} |\tilde{S}_{Q_iP_j}| + |dQ| \sum_{j=1} |\tilde{S}_{Q_iQ_j}|$$

$$\overset{(e)}{<} \frac{1}{4} (|dP| + |dQ|)$$

$$\overset{(f)}{\leq} \frac{1}{2} |dX|. \tag{6.60}$$

(a) Vectors dB and dX have $2n$ components.
(b) Eqs. (6.57) and (6.58)
(c) $|X_1, \ldots, X_{2n}| = \max(|P_1, \ldots, P_n|, |Q_1, \ldots, Q_n|)$
(d) Assume that first component is maximal; similarly for $(n+1)$-th.
(e) n sum terms smaller than $1/(4n)$ according to Eq. (6.59)
(f) $|dP| \leq |dX|$ and $|dQ| \leq |dX|$

Therefore,

$$|dB(X)| - |dX| \leq |dB(X) - dX| < \frac{1}{2} |dX|, \tag{6.61}$$

and thus

$$|dB(X)| < \frac{3}{2} |dX| < 2 |dX|, \tag{6.62}$$

which is statement (III) of the lemma.

Finally, since $|P - p| = |S_q|$ and $\|S\| \leq M$, one has $|P - p| \leq M/\delta$ by the Cauchy estimate, if $Q \in U - 3\beta - \delta$ (it suffices that $Q \in U - \delta$), which is statement (IV). □

6.8   LEMMA G4. DOMAIN SIZE OF A BOUNDED MAP

Lemma G4 *Let* $A : \overline{B}_\varepsilon(x_0) \subset R \to R$ *be a map from a closed ball* $\overline{B}_\varepsilon(x_0)$ *to* $\mathbb{R}^n$. *For all* $x \in \overline{B}_\varepsilon(x_0)$,

$$\theta \,|dx| \le |dA(x)| \le \Theta |dx| \qquad (6.63)$$

shall hold, with $0 < \theta < \Theta < \infty$. *Then*

$$\overline{B}_{\theta\varepsilon}(Ax_0) \subset A(\overline{B}_\varepsilon(x_0)) \subset \overline{B}_{\Theta\varepsilon}(Ax_0). \qquad (6.64)$$

Proof (i) We show the second subset statement first. The claim is that $|x - x_0| \le \varepsilon$ implies $|Ax - Ax_0| \le \Theta\varepsilon$. This is indeed true, as the Lagrange formula in Eq. (5.42) is applicable (see Eq. 6.63) and gives for $|x - x_0| \le \varepsilon$,

$$|Ax - Ax_0| \le \Theta|x - x_0| \le \Theta\varepsilon. \qquad (6.65)$$

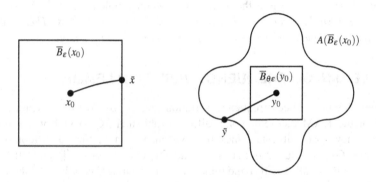

Figure 6.14   Inclusion of a ball in the A-image

(ii) The first subset statement in Eq. (6.64) is illustrated in Fig. 6.14, with $y_0 = Ax_0$. Let Y be an arbitrary point (resp. the coordinate tuple of this point) and $y(t) = y_0 + t(Y - y_0)$ with $0 \le t < \infty$ a straight ray starting at y_0 and passing through Y. For sufficiently small t, $y(t)$ lies within the A-image of $\overline{B}_\varepsilon(x_0)$,

$$y(t) \in A(\overline{B}_\varepsilon(x_0)), \qquad\qquad t \le \bar{t}, \qquad (6.66)$$

where $\bar{t}$ depends on Y. There is thus a continuous curve $x(t)$ (if A is not one-to-one, there are different branches; select any of them) with $x(0) = x_0$ and $x(\bar{t}) = \bar{x}$, where $\bar{y} = A\bar{x}$. For $0 \le t \le \bar{t}$ this defines an inverse map $x(t) = A^{-1}y(t)$ of the straight line segment from y_0 to $\bar{y}$ to $\overline{B}_\varepsilon(x_0)$. The point $\bar{x}$ lies on the boundary of $\overline{B}_\varepsilon(x_0)$, and $|\bar{x} - x_0| = \varepsilon$ for the maximum norm.

For dy along this straight segment – this is required for the Lagrange formula in Eq. (5.42) – the map A^{-1} obeys (dropping the variable t)

$$|d(A^{-1}y)| \leq \frac{1}{\theta}|dy|, \tag{6.67}$$

which is a re-formulation of $|dx| \leq \theta^{-1}|d(Ax)|$ from Eq. (6.63). Applying Eq. (5.42) with $a \equiv y_0$ and $b \equiv \bar{y}$ to A^{-1} gives

$$\varepsilon = |\bar{x} - x_0| \leq \frac{1}{\theta}|\bar{y} - y_0|, \tag{6.68}$$

and so

$$|\bar{y} - y_0| \geq \theta\,\varepsilon. \tag{6.69}$$

Since Y was arbitrary, this shows that $|\theta\,\varepsilon| \leq |\bar{y} - y_0|$ for all $\bar{y}(Y)$. This implies that $\bar{B}_{\theta\varepsilon}(y_0)$ lies within $A(\bar{B}_\varepsilon(x_0))$, as was to be shown. (Note that for non-convex image domains $U = A(\bar{B}_\varepsilon(x_0))$, the $\bar{y}$ will only reach parts of ∂U.)   □

Remark The following is a theorem from general topology: *Let $f : X \to Y$ be a homeomorphism between topological spaces X and Y, and $U \subset X$. Then $f(\partial U) = \partial(fU)$, i.e., the image of the boundary is the boundary of the image.*

6.9   LEMMA G5. FREQUENCY VARIATION LEMMA

The next lemma, called 'frequency variation lemma' by Arnold, is central to the proof of the KAM theorem. For a diffeomorphism $A : G \to \Omega$ from momentum to frequency space, it establishes the existence of a perturbed diffeomorphism $A' : G_1 \to \Omega_1$ on reduced domains G_1 and Ω_1, if both $\|A - A'\|$ and $\|d(A - A')\|$ are sufficiently small, and if A and thus A' obey the standard condition of the implicit function theorem that their Jacobi matrices have nonvanishing determinant.

Lemma G5 *Let $A : G \to \Omega$ with $A(p) = \omega$ be a diffeomorphism of a domain G in momentum space onto the domain Ω in frequency space (i.e., $AG = \Omega$, A is surjective) such that for all $p \in G$,*

$$\theta\,|dp| \leq |dA(p)| \leq \Theta|dp|. \tag{6.70}$$

Let $A' : G - \beta \to \Omega$ with $\beta > 0$ be a 'perturbed' map[8] *with $A'(p) = A(p) + \Delta(p)$, subject to the bounds*

$$|\Delta(p)| < \beta, \tag{6.71}$$

$$|d\Delta(p)| < \delta\theta\,|dp|, \tag{6.72}$$

where $0 < \delta < 1$ and $\theta > 0$. Then the following holds:

---

[8] The restriction to $G - \beta$ is due to the fact that $|p - P| < \beta$, see Eq. (4.77). With domain G for $A = A(p)$, the map $A' = A(P)$ should be restricted to $G - \beta$, in order to avoid P-values for which A' is not defined.

(I) *There exist domains G_1 and G' such that*

$$G_1 \subset G_1 + b \subset G' \subset G - \beta \subset G. \tag{6.73}$$

(II) $A' : G' \to \Omega'$ *is a diffeomorphism of G' onto Ω', and* [9]

$$(1 - \delta)\,\theta\,|dp| < |dA'(p)| < (1 + \delta)\,\Theta\,|dp|. \tag{6.74}$$

(III) $A'G_1 = \Omega_1 = \Omega_0 - d$ *with* $d = 2\Theta b + (5 + \Theta)\beta$ *and* $A'(G_1 + b) \subset \Omega_0$.

(IV) *The measure of the 'lost' domain $G \backslash G_1$ is given by*

$$\text{mes}\,(G \backslash G_1) \le \theta^{-n}\,\text{mes}\,(\Omega \backslash (\Omega_0 - d - \beta)). \tag{6.75}$$

Here $\Omega_0 \subset \Omega$ must be sufficiently large and $b > 0$ sufficiently small so that $\Omega_d - d - \beta$ in (IV) is non-empty; apart from this, Ω_0 and b are arbitrary.

Proof The sequence of G-domains in (I) is constructed in steps (i) to (iii). Statements (I) to (III) are proved at the end of (iii) and (IV) is proved in (iv).

(i) *Construction of G''* We consider first the map $A' \circ A^{-1}$ and apply Lemmas G1 and G2. Let $p \in G - \beta$ and $\omega = A(p)$. Then from $A'(p) = A(p) + \Delta(p)$,

$$A'(A^{-1}(\omega)) - A(A^{-1}(\omega)) = \Delta(p), \tag{6.76}$$

and since $|\Delta(p)| < \beta$,

$$\|A' \circ A^{-1} - E\| < \beta. \tag{6.77}$$

Furthermore, for the total differential,

$$|d(A' \circ A^{-1}(\omega))| = |dA'(p)| = |dA(p) + d\Delta(p)| \ge \big| \, |dA(p)| - |d\Delta(p)| \, \big|$$

$$\overset{(a)}{\ge} \big| \, |d\omega| - |d\Delta(p)| \, \big| \overset{(b)}{>} \big| \, |d\omega| - \delta\theta|dp| \, \big| \overset{(c)}{\ge} (1 - \delta)\,|d\omega|. \tag{6.78}$$

(a) since $dA = d\omega$
(b) since $|d\Delta(p)| < \delta\theta|dp|$
(c) since $\theta|dp| \le |dA|$ and $0 < \delta < 1$

Therefore,

$$|d\omega| \ne 0 \to |d(A' \circ A^{-1}(\omega))| \ne 0. \tag{6.79}$$

According to Lemma G2, Eqs. (6.77) and (6.79) together with continuity of $A' \circ A^{-1}$ imply that

$$A' \circ A^{-1} : A(G - \beta) - 4\beta \to \Omega \tag{6.80}$$

---

[9]Note '$\le$' in Eq. (6.70) vs. '$<$' in Eq. (6.74). We mostly retain Arnold's choice, instead of attempting unification.

is a diffeomorphism. Define the (auxiliary) domain G'' by

$$G'' = A^{-1}[A(G-\beta) - 4\beta]. \tag{6.81}$$

Then, since A (by assumption) and $A' \circ A^{-1}$ (by Eq. 6.80) are diffeomorphisms, also

$$A' : G'' \to \Omega \tag{6.82}$$

is a diffeomorphism of the domain G'' (and of any subdomain $G' \subset G''$). From Eq. (6.77), the Arnold-Brouwer Lemma G1, i.e., $U - \varepsilon \subset AU$ if $|Ax - x| \leq \varepsilon$ gives, replacing A in this lemma by $A' \circ A^{-1}$, ε by β and U by $A(G-\beta) - 4\beta$, that

$$A(G-\beta) - 5\beta \subset (A' \circ A^{-1})[A(G-\beta) - 4\beta] = A'G''. \tag{6.83}$$

For later use, we also note that, with $a' \subset c' \to f^{-1}(a') \subset f^{-1}(c')$ for any function f and sets a', c',

$$G'' = A^{-1}[A(G-\beta) - 4\beta] \subset A^{-1}[A(G-\beta)] = G - \beta. \tag{6.84}$$

From Eqs. (6.70) and (6.72) one has, with $0 < \delta < 1$ and for all p,

$$|dA'(p)| = |d(A+\Delta)| \leq |dA(p)| + |d\Delta(p)|$$
$$< \Theta|dp| + \delta\theta|dp| \leq (1+\delta)\Theta|dp|, \tag{6.85}$$

and

$$|dA'(p)| \geq \big| |dA(p)| - |d\Delta(p)| \big| > (1-\delta)\theta|dp|. \tag{6.86}$$

The following items (ii) and (iii) are covered in exactly three lines in Arnold (1963a). We quote them in full so the reader can decide whether it is worth to work through (ii) and (iii).

> But by 4°. [Lemma G4] $A(G-\beta) \supseteq \Omega - \Theta\beta$, hence $A'G'' \supseteq \Omega' = \Omega - (5+\Theta)\beta$. Writing $G' = A'^{-1}\Omega'$, $G_1 = A'^{-1}\Omega_1$, where $\Omega_1 = \Omega_0 - d$, $d = 2\Theta b + (5+\Theta)\beta$ we have $\Omega_1 + 2\Theta b \subseteq \Omega'$ and by 4°. $G_1 + b \subseteq G'$. The conclusions (II) and (III) are now evident. (Arnold 1963a, p. 28)

(ii) *Construction of Ω' and G'* Let

$$S = G \backslash (G-\beta),$$
$$T = \Omega \backslash (\Omega - \Theta\beta) \tag{6.87}$$

be strips of (maximum) width β and $\Theta\beta$ taken off from G and Ω, respectively. Since $A : G \to \Omega$ is a diffeomorphism, the image of the boundary is the boundary of the image,

$$A(\partial G) = \partial\Omega. \tag{6.88}$$

Therefore, and since $f(a \cup c) = f(a) \cup f(c)$ for any function f and sets a, c, one has

$$A(S) \overset{(a)}{=} A\left(\bigcup_{p \in \partial G} B_\beta(p) \right) = \bigcup_{p \in \partial G} A(B_\beta(p)) \overset{(b)}{\subset} \bigcup_{\omega \in \partial \Omega} B_{\Theta\beta}(\omega) = T. \tag{6.89}$$

In (a), Eq. (2.20) is used for S written as union of boundary balls of radius β and in (b), Eq. (6.64) from Lemma G4 is used,[10] which applies to A by assumption (6.70). Furthermore, Eq. (6.88) is used in (b) to write the union along ∂G as union along $\partial \Omega$. Using $f(a) \backslash f(c) \subset f(a \backslash c)$ for any function f and sets a, c for which $c \subset a$, one obtains

$$\begin{aligned}
\Omega - \Theta\beta &= \Omega \backslash T \\
&\subset \Omega \backslash A(S) \\
&= A(G) \backslash A(G \backslash (G - \beta)) \\
&\subset A(G \backslash [G \backslash (G - \beta)]) \\
&= A(G - \beta),
\end{aligned} \tag{6.90}$$

which is also intuitively clear. Introduce now

$$\Omega' = \Omega - (5 + \Theta)\beta. \tag{6.91}$$

Then

$$\Omega' = \Omega - \Theta\beta - 5\beta \overset{(6.90)}{\subset} A(G - \beta) - 5\beta \overset{(6.83)}{\subset} A'G''. \tag{6.92}$$

Finally, let

$$G' = A'^{-1} \Omega'. \tag{6.93}$$

Using again $a' \subset c' \rightarrow f^{-1}(a') \subset f^{-1}(c')$, this gives

$$G' \overset{(6.92)}{\subset} A'^{-1}A' G'' = G''. \tag{6.94}$$

(iii) *Construction of Ω_1 and G_1* Let $b > 0$ be a small number and

$$d = 2\Theta b + (5 + \Theta)\beta. \tag{6.95}$$

For $\Omega_0 \subset \Omega$ let

$$\Omega_1 = \Omega_0 - d. \tag{6.96}$$

Here Ω_0 must be sufficiently large and b sufficiently small[11] so that Ω_1 (actually $\Omega_1 - \beta$, see Eq. 6.106) is non-empty; apart from this, Ω_0 and b are arbitrary. Since $(\Omega - \delta) + \delta \subset \Omega$ for any $\delta > 0$ (see Eq. 2.23) one obtains

$$\Omega_1 + 2\Theta b \subset \Omega_0 - (5 + \Theta)\beta \subset \Omega - (5 + \Theta)\beta \overset{(6.91)}{=} \Omega'. \tag{6.97}$$

---

[10]Lemma G4 is for closed balls, whereas the balls in Eq. (6.89) are open. This poses no problem.

[11]In the proof of the Fundamental theorem, a boundary strip of width $b = 3\beta$ is removed from the momentum domain G to allow for a canonical transformation of the Hamiltonian. Ω_0 is the frequency domain Ω with resonance strips removed using a Diophantine condition.

Let then

$$G_1 = A'^{-1}\, \Omega_1. \tag{6.98}$$

With this, the construction of domains is complete. We prove two subset relations for G_1. First, from Eq. (6.85) with $\delta < 1$, using Eq. (6.64) of Lemma G4 and writing now $\omega = A'(p)$ instead of $\omega = A(p)$ before,

$$A'(\bar{B}_b(p)) \subset \bar{B}_{2\Theta b}(\omega), \tag{6.99}$$

and thus

$$\bar{B}_b(p) \subset A'^{-1}(\bar{B}_{2\Theta b}(\omega)). \tag{6.100}$$

From this, with $\partial G_1 = A'^{-1}(\partial \Omega_1)$ (A' is a diffeomorphism) and $f^{-1}(a \cup c) = f^{-1}(a) \cup f^{-1}(c)$,

$$
\begin{aligned}
G_1 + b \overset{(2.22)}{=}\;& G_1 \cup \bigcup_{p \in \partial G_1} \bar{B}_b(p) \\[1ex]
\overset{(6.100)}{\subset}\;& A'^{-1}\, \Omega_1 \cup \bigcup_{\omega \in \partial \Omega_1} A'^{-1}(\bar{B}_{2\Theta b}(\omega)) \\[1ex]
=\;& A'^{-1}\!\left(\Omega_1 \cup \bigcup_{\omega \in \partial \Omega_1} \bar{B}_{2\Theta b}(\omega) \right) \\[1ex]
\overset{(2.22)}{=}\;& A'^{-1}(\Omega_1 + 2\Theta b) \\[1ex]
\overset{(6.97)}{\subset}\;& A'^{-1}\, \Omega' \\[1ex]
\overset{(6.93)}{=}\;& G'.
\end{aligned} \tag{6.101}
$$

The second subset relation for G_1 is derived more easily,

$$A'(G_1 + b) \overset{(6.101)}{\subset} \Omega_1 + 2\Theta b \overset{(6.97)}{\subset} \Omega_0 - (5 + \Theta)\beta \subset \Omega_0. \tag{6.102}$$

To summarize the results so far, we have obtained the following sequences of subdomains. In general (but not necessarily) $A'G'' \not\subset A(G - \beta) - 4\beta$.

$$
\begin{array}{ccccccccc}
\Omega_1 & \overset{(a)}{\subset} & A'(G_1 + b) & \overset{(b)}{\subset} & \Omega' & \overset{(c)}{\subset} & A(G - \beta) - 5\beta & \overset{(c)}{\subset} & A'G'' \not\subset A(G - \beta) - 4\beta \subset A(G - \beta) \overset{(d)}{\subset} \Omega \\
\Big\downarrow A'^{-1} & & & & \Big\downarrow A'^{-1} & & & & \quad \Big\uparrow A' \qquad \overset{(f)}{\;}\Big\downarrow A^{-1} \qquad \Big\uparrow A \qquad \Big\uparrow A \\
G_1 \subset & & G_1 + b \underset{(b)}{\subset} G' & & & & \underset{(e)}{\subset} \; G'' = & & G'' \underset{(g)}{\subset} G - \beta \subset G
\end{array}
$$

(a): (6.98), (b): (6.101), (c): (6.92), (d): $\Omega = A(G)$, (e): (6.94), (f): (6.81), (g): (6.84)

Note that only Ω_1 and G_1 (and thus also $G_1 + b$ and its image by A') depend on Ω_0. We have thus proved statements (I) to (III) of the lemma:

(I) The sequence $G_1 \subset G_1 + b \subset G' \subset G - \beta \subset G$ is part of the lower sequence.

(II) From Eq. (6.82), $A' : G'' \to A'G''$ and thus also $A' : G' \to \Omega'$ with $G' \subset G''$ are diffeomorphisms. Equations (6.85) and (6.86) are the estimates on $|dA'(p)|$.

(III) $A'G_1 = \Omega_1$ by Eq. (6.98) and $A'(G_1 + b) \subset \Omega_0$ by Eq. (6.102).

(iv) Finally, to show statement (IV), consider the diffeomorphism $A' : G_1 \to \Omega_1$ with

$$A(p) - A'(p) = -\Delta(p). \tag{6.103}$$

The bound on $\Delta(p)$ in Eq. (6.71) gives, with $\omega = A'(p)$,

$$|(A \circ A'^{-1})(\omega) - \omega| \le \beta, \tag{6.104}$$

where $A \circ A'^{-1} : \Omega_1 \to \Omega$. Therefore,

$$\Omega_1 - \beta \overset{(6.13)}{\subset} (A \circ A'^{-1})(\Omega_1) \overset{(6.98)}{=} AG_1, \tag{6.105}$$

where the subset relation is from the Arnold-Brouwer Lemma G1. Using $|dA| \ge \theta \, |dp|$, and since G and Ω have dimension n, one has

$$\text{mes}\,(G \backslash G_1) = \int_{G \backslash G_1} dp = \int_{\Omega \backslash AG_1} d\omega \; \frac{1}{\det(\partial A_i / \partial p_j)}$$

$$\overset{(6.70)}{\le} \theta^{-n} \,\text{mes}\,(\Omega \backslash AG_1) \overset{(6.105)}{\le} \theta^{-n} \,\text{mes}\,(\Omega \backslash (\Omega_1 - \beta)), \tag{6.106}$$

which is statement (IV) of the lemma.   □

Remark We summarize the steps that lead to the somewhat awkward number $d = 2\Theta b + (5 + \Theta)\beta$, to show that the large domain reduction $\Omega - d$ is indeed necessary to make A' a diffeomorphism and, furthermore, to ensure a condition from the Fundamental theorem of Section 4.1. We argue for Ω, which can be replaced by its subset Ω_0 (resonance strip removed).

(i) From Eq. (6.90),

$$\Omega - \Theta\beta \subset A(G - \beta), \tag{6.107}$$

which is a direct consequence of $|dA(p)| \le \Theta \, |dp|$.

(ii) Subtracting in Eq. (6.107) 5β on both sides one obtains as in Eq. (6.83), applying the Arnold-Brouwer Lemma G1 to the near-identity $\|A' \circ A^{-1} - E\| < \beta$,

$$A'G' = \Omega' = \Omega - (5 + \Theta)\beta \subset A(G - \beta) - 5\beta \subset A'G'', \tag{6.108}$$

where

$$G'' = A^{-1}(A(G - \beta) - 4\beta). \tag{6.109}$$

Thus one β in the above 5β is due to Lemma G1.

(iii) The remaining 4β originate in Lemma G2 (see the sentence after Eq. 6.30), and ensure that $A' \circ A^{-1}$ is a diffeomorphism of $A(G - \beta) - 4\beta$; and thus that A' is a diffeomorphism of $A^{-1}(A(G - \beta) - 4\beta) = G''$ (since A is a diffeomorphism), and thus also on $G' \subset G''$. Thus A'^{-1} gives a *unique* momentum value P for $\omega \in \Omega'$.

(iv) The final reduction of Ω' to

$$\Omega_1 = \Omega' - 2\Theta b = \Omega - 2\Theta b - (5 + \Theta)\beta \qquad (6.110)$$

and $G_1 = A'^{-1}\Omega_1$ originates from the Fundamental theorem. This requires a reduction of the momentum domain by a boundary layer of width 3β. Let thus $b = 3\beta$. Then Eq. (6.64) from Lemma G4 gives, since $|dA'(P)| < (1 + \delta)\Theta\,|dP| < 2\Theta\,|dP|$, that

$$A'(B_b(P)) \subset B_{2\Theta b}(\omega). \qquad (6.111)$$

To ensure, therefore, that the distance of P from the boundary of the momentum domain is $> b$, one has to take another $2\Theta b$ off the frequency domain Ω'. (Since $\Theta > 1$ and b are arbitrary, this reduction cannot be assumed to be contained in $(5 + \Theta)\beta$.) This gives Eq. (6.110) for Ω_1 and (see Eq. 6.101)

$$G_1 + b \subset G' = A'^{-1}\Omega', \qquad (6.112)$$

as is required by the Fundamental theorem (and A'^{-1} is unique on Ω').

7 Convergence Lemmas

The four Lemmas C1 to C4 are proved in section 4.4 of Arnold (1963a). They address the infinite sequence of canonical transformations that shift perturbations to higher orders.

7.1 LEMMA C1. ITERATED CANONICAL TRANSFORMATION

Lemma C1 *Let (F_s) with $s = 0, 1, 2, \ldots$ be a sequence of shrinking domains and (B_s) with $B_s : F_s \to F_{s-1}$ a sequence of diffeomorphisms such that (with $c > 0$ an arbitrary constant):*

$\qquad$ (A) $\|B_s - E\| \le d_s$,
$\qquad$ (B) $F_s \subset F_{s-1} - d_s$,
$\qquad$ (C) $\|dB_s\| \le 2|dx|$,
$\qquad$ (D) $d_s \le c\, 4^{-s}$.

Then the sequence (S_s) of maps

$$S_s = B_1 \circ B_2 \circ \cdots \circ B_s, \qquad\qquad s = 1, 2, 3, \ldots \qquad (7.1)$$

converges uniformly to a continuous map S_∞ with domain $F_\infty = \bigcap F_s$ such that $|S_\infty - E| < c$.

Proof Let $x \in F_s$. Then $|B_s(x) - x| \le d_s$ by (A). Therefore by (B), both $x \in F_{s-1}$ and $B_s(x) \in F_{s-1}$. Both B_{s-1} and S_{s-1} are diffeomorphisms on F_{s-1}. We use (C) to apply the Lagrange estimate from Lemma A4 to S_{s-1}. With $x = (p, q)$ of length $2n$, one has

$$|dB_s(x)| = \max_{i \in \{1, \ldots, 2n\}} \left(\left| \sum_{j=1}^{2n} \frac{\partial B_{s,i}}{\partial x_j} \, dx_j \right| \right) \le 2|dx|. \qquad (7.2)$$

Using $(dx_1, \ldots, dx_{2n}) = (\pm 1, \ldots, \pm 1)\, dx$, where each component is independently $+1$ or -1 (cf. the reasoning for the matrix norm in Eq. 2.6), gives

$$\max_{i \in \{1, \ldots, 2n\}} \left(\sum_{j=1}^{2n} \left| \frac{\partial B_{s,i}}{\partial x_j} \right| \right) \le 2, \qquad (7.3)$$

and correspondingly for B_1 to B_{s-1}. From the chain rule $d(f \circ g)(x) = f'g'\, dx$ and its

DOI: 10.1201/9781003287803-7

generalization to vector-valued functions one has for arbitrary $x \in F_{s-1}$,

$$
\begin{aligned}
|dS_{s-1}(x)| &= |d(B_1 \circ B_2 \circ \cdots \circ B_{s-1})(x)| \\
&= \max_I \left(\left| \sum_i \frac{\partial B_{1,I}}{\partial x_i} \sum_j \frac{\partial B_{2,i}}{\partial x_j} \cdots \sum_l \frac{\partial B_{s-1,k}}{\partial x_l} \, dx_l \right| \right) \\
&\leq \max_I \left(\sum_j \left| \frac{\partial B_{1,I}}{\partial x_i} \right| \right) \max_J \left(\sum_j \left| \frac{\partial B_{2,J}}{\partial x_j} \right| \right) \cdots \max_L \left(\sum_l \left| \frac{\partial B_{s-1,L}}{\partial x_l} \right| \right) |dx| \\
&\leq 2^{s-1} |dx|,
\end{aligned}
\tag{7.4}
$$

where $i, I, j, J, \ldots, l, L$ are independent indices running from 1 to $2n$. By Eq. (7.4), the Lagrange estimate in Eq. (5.42) can be applied to S_{s-1} along the segment $xB_s(x) \subset F_{s-1}$ of length $\leq d_s$, which gives with assumption (D),

$$
|S_s(x) - S_{s-1}(x)| = |S_{s-1}(B_s(x)) - S_{s-1}(x)| \leq 2^{s-1} \, d_s \leq c \frac{2^{s-1}}{4^s}, \tag{7.5}
$$

or

$$
\|S_s - S_{s-1}\| \leq \frac{c}{2^{s+1}}. \tag{7.6}
$$

The sequence (S_s) of functions S_s is therefore a Cauchy sequence independent of x, and thus converges uniformly to a function S_∞ on a domain F_∞. To check that the latter is non-empty, note that

$$
\begin{aligned}
F_1 &\subset F_0 - d_1, \\
F_2 &\subset F_1 - d_2 \subset F_0 - d_1 - d_2,
\end{aligned}
\tag{7.7}
$$

and so forth, and therefore

$$
F_\infty \subset F_0 - \sum_{s=1}^{\infty} d_s. \tag{7.8}
$$

Again with assumption (D), one has

$$
\sum_{s=1}^{\infty} d_s \leq c \left(\sum_{s=0}^{\infty} 4^{-s} - 1 \right) = c \left(\frac{1}{1 - \frac{1}{4}} - 1 \right) = \frac{c}{3}, \tag{7.9}
$$

so that F_0 must have diameter $> 2c/3$. Finally,

$$
\begin{aligned}
\|S_\infty - E\| &= \| \cdots + S_s - S_{s-1} + S_{s-1} - S_{s-2} + \cdots - S_1 + B_1 - E \| \\
&\leq \sum_{s=2}^{\infty} \|S_s - S_{s-1}\| + \|B_1 - E\| \\
&\leq \frac{c}{2} \sum_{s=2}^{\infty} 2^{-s} + \frac{c}{4} = \frac{c}{2} \sum_{s=1}^{\infty} 2^{-s} = \frac{c}{2} < c,
\end{aligned}
\tag{7.10}
$$

which proves the lemma. □

7.2 LEMMA C2. INTEGRAL CURVES AND SEGMENTS

The next lemma gives an estimate for the short-time deviation of the integral curve of a vector field starting in x_0 from a straight line through x_0.

Lemma C2 *Let F be a d-neighborhood of the straight line segment (point x_0, constant vector Y)*

$$x = x_0 + Yt, \qquad\qquad 0 \le t \le \frac{d}{\varepsilon}. \qquad (7.11)$$

Let [1] X_x be a smooth vector field on F, and

$$|X_x - Y| \le \varepsilon \qquad (7.12)$$

in F. Let x(t) be the solution of the differential equation

$$\frac{dx}{dt} = X_x \qquad (7.13)$$

with initial condition $x(0) = x_0$. Then (see Fig. 7.1)

$$|x(t) - (x_0 + Yt)| \le d \qquad for \qquad 0 \le t \le \frac{d}{\varepsilon}. \qquad (7.14)$$

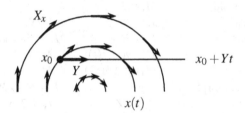

Figure 7.1 Integral curves $x(t)$ of vector field X_x and straight-line approximation $x_0 + Yt$

Proof Let

$$y(t) = x(t) - (x_0 + Yt). \qquad (7.15)$$

Then $y(0) = x(0) - x_0 = 0$. Thus there exists a $t_0 > 0$ with

$$|y(t)| < d \quad for \ \ 0 \le t < t_0 \qquad and \qquad |y(t_0)| = d. \qquad (7.16)$$

Furthermore,

$$\left| \frac{dy}{dt} \right| = \left| \frac{dx}{dt} - Y \right| = |X_x - Y| \le \varepsilon. \qquad (7.17)$$

---

[1] To reduce bracketing in the next section, we write X_x instead of $X(x)$, and f_*X instead of $f_*(X)$ (see below).

The Lagrange estimate in Eqs. (5.41) and (5.42) gives for $y(t)$, with t on the straight segment $[0, t_0]$,

$$|dy| \leq \varepsilon \, dt \quad \rightarrow \quad |y(t_0) - y(0)| \leq \varepsilon t_0, \tag{7.18}$$

or

$$t_0 \geq \frac{|y(t_0)|}{\varepsilon} = \frac{d}{\varepsilon}. \tag{7.19}$$

Thus generally, inserting Eq. (7.15) for $y(t)$ in Eq. (7.16),

$$|x(t) - (x_0 + Yt)| \leq d \qquad \text{for} \qquad 0 \leq t \leq \frac{d}{\varepsilon}, \tag{7.20}$$

which is Eq. (7.14).   □

7.3   THE PUSH FORWARD

As a preparation for Lemma C3, we include here a section on some basic differential geometry, the so-called *push forward* (a re-formulation of the Jacobi matrix) and its relation to flows induced by vector fields. The symbols used are collected in the following list.

Object	Symbol	Property
Manifold	M, N, P	
Point	x, y, z	$\in M, N, P$
Function	F, G, H	$M, N, P \rightarrow \mathbb{R}$
Curve	$x(t), y(t), z(t)$	$\subset M, N, P$
Flow	$\phi^t(x_0), \psi^t(y_0)$	$\subset M, N$
Diffeomorphism	f, g	$M \xrightarrow{f} N \xrightarrow{g} P$
Vector	X_x, Y_y, Z_z	$\in T_x M, T_y N, T_z P$

The vector fields X and Y shall generate flows $\phi^t(x_0)$ and $\psi^t(x_0)$ (instead of ϕ^t_X and ϕ^t_Y in Section 1.7). The tangent vector X_{x_0} to a curve $x(t)$ in $\mathbb{R}^m$ in the point $x_0 = x(0)$ is given by Eq. (1.153),

$$X_{x_0} = \frac{d}{dt}\bigg|_{t=0} x(t). \tag{7.21}$$

Let $f : \mathbb{R}^m \rightarrow \mathbb{R}^n$ be a differentiable map. Then the *push forward* f_* of X is defined by

$$(f_* X)_{f(x_0)} = \frac{d}{dt}\bigg|_{t=0} f(x(t)). \tag{7.22}$$

Definition (see Kobayashi & Nomizu 1963, p. 8)   If the vector X_{x_0} is tangent to the curve $x(t)$ in x_0 ($\in \mathbb{R}^m$ or M), then the *push forward* $(f_* X)_{f(x_0)}$ is tangent to the curve $f(x(t))$ in $f(x_0)$ ($\in \mathbb{R}^n$ or N). Or, f_* maps the tangential vector of a curve in a point to the tangential vector of the image curve in the image point, see Fig. 7.2. Or, the diagram in Fig. 7.3 is *commutative*.

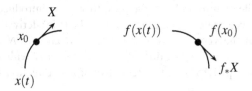

Figure 7.2   Definition of f_*

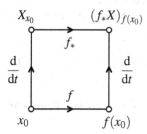

Figure 7.3   Commutative diagram for f_*

In Eq. (7.22), the reference to the image point $f(x_0)$ is not quite obvious. Indeed, in local coordinates, f_* is the Jacobi matrix, and is calculated at x_0, but used to calculate a vector at $f(x_0)$. This is made explicit in Spivak[2] (1970), p. 3-3, who introduces f_* by (symbols adapted)

$$(f_*X)_{f(x_0)} = [(Df)(x_0)(X)]_{f(x_0)}, \tag{7.23}$$

where $Df : \mathbb{R}^m \to \mathbb{R}^n$ is the linearization of f, i.e., the $n \times m$ Jacobi matrix $(\partial f_i / \partial x_j)$. The chain rule applied to Eq. (7.22) gives

$$(f_*X)_{f(x_0)} = (Df)(x_0)(X_{x_0}), \tag{7.24}$$

and one may expect – erroneously – from the right side a subscript x_0 in Eqs. (7.23) and (7.24). However, if $f : M \to N$ between manifolds M and N (see below), then $(f_*)_x$ is a map from $T_x M$ to $T_{f(x)} N$. Thus the correct correspondence is (see Kobayashi & Nomizu 1963, p. 10, and the figures above),

$$(f_*X)_{f(x_0)} = (f_*)_{x_0}(X_{x_0}). \tag{7.25}$$

We will therefore take great care in the following to make very explicit the 'foot points' of all vectors, i.e., the tangent spaces to which they belong.

---

[2]Spivak speaks of the 'apparently anomalous features' of f_*.

On differentiable manifolds M, N, the push forward is introduced more generally via the action of a vector field on a function, i.e., the (Lie) derivative of the function in the direction of the vector field. Let $F : M \rightarrow \mathbb{R}$ and $G : N \rightarrow \mathbb{R}$ be arbitrary differentiable functions and $x(t)$ a differentiable curve in M with tangent vector X_{x_0} in $x_0 = x(0)$. Then Eq. (1.154) for the derivative of F in direction X_{x_0} is

$$X_{x_0} F = \left. \frac{d}{dt} \right|_{t=0} F(x(t)) \qquad (7.26)$$

Definition Let $f : M \rightarrow N$ be a diffeomorphism between manifolds M and N. Then the *push forward* $f_* : TM \rightarrow TN$ between the tangent bundles is defined by

$$(f_* X)_{f(x_0)} G = \left. \frac{d}{dt} \right|_{t=0} G(f(x(t))) \qquad (7.27)$$

The following lemma combines the two definitions and shows that the definition of f_* is very natural.

Lemma *With manifolds M, N, vector field $X \in TM$, diffeomorphism $f : M \rightarrow N$ and differentiable function $G : N \rightarrow \mathbb{R}$, one has*

$$(f_* X) G = X(G \circ f). \qquad (7.28)$$

Proof Let X_{x_0} be tangent vector to the curve $x(t)$ in $x_0 = x(0)$. Then

$$(f_* X)_{f(x_0)} G \overset{(7.27)}{=} \left. \frac{d}{dt} \right|_{t=0} G(f(x(t))) = \left. \frac{d}{dt} \right|_{t=0} (G \circ f)(x(t)) \overset{(7.26)}{=} X_{x_0}(G \circ f).$$

The next lemma states that $*$ is a functor, see Abraham & Marsden (1978), p. 21. For a shorter proof using the chain rule, see Spivak (1970), p. 3-4. Figure 7.4 shows the situation.

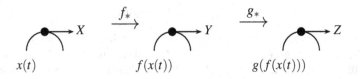

Figure 7.4 Composition of f_* and g_*

Lemma *Let $f : M \rightarrow N$ and $g : N \rightarrow P$ be diffeomorphisms between manifolds M, N, P. Then*

$$(g \circ f)_* = g_* \circ f_*. \qquad (7.29)$$

Proof Let again X_{x_0} be tangent to $x(t)$ in x_0. Let $H : P \to \mathbb{R}$ and $y(t) = f(x(t))$ with $y(0) = y_0 = f(x_0)$, and abbreviate[3] $Y = f_* X$ and $Z = g_* Y$. Then

$$Y_{y_0} = (f_*)_{x_0} X_{x_0}. \tag{7.30}$$

With this, one has, using repeatedly Eqs. (7.25) and (7.27),

$$
\begin{aligned}
(((g \circ f)_*)_{x_0} X_{x_0}) H &= ((g \circ f)_* X)_{(g \circ f)(x_0)} H \\
&= \left. \frac{\mathrm{d}}{\mathrm{d}t} \right|_{t=0} H((g \circ f)(x(t))) \\
&= \left. \frac{\mathrm{d}}{\mathrm{d}t} \right|_{t=0} H(g(y(t))) \\
&= (g_* Y)_{g(y_0)} H \\
&= ((g_*)_{y_0} Y_{y_0}) H \\
&= ((g_*)_{y_0} ((f_*)_{x_0} X_{x_0})) H \\
&= ((g_*)_{y_0} (f_* X)_{y_0}) H \\
&= (g_* (f_* X))_{g(y_0)} H \\
&= ((g_* \circ f_*) X)_{g(f(x_0))} H \\
&= ((g_* \circ f_*)_{x_0} X_{x_0}) H, \tag{7.31}
\end{aligned}
$$

and therefore

$$((g \circ f)_*)_{x_0} = (g_* \circ f_*)_{x_0}, \tag{7.32}$$

as was to be shown.   □

The following theorem is the main result of this section, and is used in Lemma C3.

Theorem *Let $f : M \to N$ be a diffeomorphism between manifolds M and N, and X a vector field on M that generates the flow ϕ^t. Then $f_* X$ generates the flow $\psi^t = f \circ \phi^t \circ f^{-1}$ on N.*

Proof (i) First, we show that $f \circ \phi^t \circ f^{-1}$ is a flow if ϕ^t is a flow.

$$
\begin{aligned}
\psi^0 &= f \circ \phi^0 \circ f^{-1} = f \circ f^{-1} = E, \\
\psi^s \circ \psi^t &= (f \circ \phi^s \circ f^{-1}) \circ (f \circ \phi^t \circ f^{-1}) = f \circ \phi^s \circ \phi^t \circ f^{-1} = f \circ \phi^{s+t} \circ f^{-1} = \psi^{s+t}, \\
(\psi^t)^{-1} &= (f \circ \phi^t \circ f^{-1})^{-1} = (f^{-1})^{-1} \circ (\phi^t)^{-1} \circ f^{-1} = f \circ \phi^{-t} \circ f^{-1} = \psi^{-t}.
\end{aligned}
\tag{7.33}
$$

---

[3] The variable Y has a different meaning in the present section from that in the foregoing and next section.

Second, we show that f_*X generates $f \circ \phi \circ f^{-1}$ (see Spivak 1970, p. 5-34). Equation (7.26) is

$$X_{x_0} F = \frac{d}{dt}\Big|_{t=0} F(x(t)), \qquad x(0) = x_0. \tag{7.34}$$

Written with the flow $x(t) = \phi^t(x_0)$, this becomes

$$X_{x_0} F = \frac{d}{dt}\Big|_{t=0} F(\phi^t(x_0)), \qquad \phi^0(x_0) = x_0. \tag{7.35}$$

Let $Y = f_*X$ generate $y(t)$,

$$Y_{y_0} G = \frac{d}{dt}\Big|_{t=0} G(y(t)), \qquad y(0) = y_0. \tag{7.36}$$

This is rewritten with $y(t) = \psi^t(y_0)$, where $y = f(x)$ and $y_0 = f(x_0)$, as

$$
\begin{aligned}
Y_{y_0} G &= \frac{d}{dt}\Big|_{t=0} G(f(x(t))) \\
&= \frac{d}{dt}\Big|_{t=0} G(f \circ \phi^t(x_0)) \\
&= \frac{d}{dt}\Big|_{t=0} G((f \circ \phi^t \circ f^{-1})(y_0)), \qquad (f \circ \phi^0 \circ f^{-1})(y_0) = y_0, \tag{7.37}
\end{aligned}
$$

showing that $f \circ \phi^t \circ f^{-1}$ is the flow of f_*X.

Kobayashi & Nomizu (1963), p. 14, give a probably more direct argument for this: *X generates the flow ϕ^t* means that for any x_0,

(a) X_{x_0} is tangent to $\phi^t(x_0)$ for $t = 0$ (i.e., in $x_0 = \phi^0(x_0)$).

On the other hand, by definition of f_*,

(b) $(f_*X)_{y_0}$ is tangent to $f \circ \phi^t(x_0)$ in $y_0 = f(x_0)$.

To make (b) resemble (a), one has to find a flow ψ^t of f_*X with $y_0 = \psi^0(y_0)$: to define a tangent in y_0, one either needs a curve $y(t)$ through y_0, conventionally $y_0 = y(0)$, or a flow $\psi^t(y_0)$ with $y_0 = \psi^0(y_0)$. Using $x_0 = f^{-1}(y_0)$, (b) becomes

(c) $(f_*X)_{y_0}$ is tangent to $f \circ \phi^t \circ f^{-1}(y_0)$ for $t = 0$ (i.e., in $y_0 = \psi^0(y_0)$).

This means (cf. item (a)) that f_*X generates the flow $f \circ \phi^t \circ f^{-1}$. $\square$

Remark Without reference to vector fields X and their transforms f_*X, the origin of the relation

$$\psi^t = f \circ \phi^t \circ f^{-1} \tag{7.38}$$

solely in terms of flows is rather obvious. Indeed, from Fig. 7.5,

$$\psi^t(y_0) = y_1 = f(x_1) = f(\phi^t(x_0)) = f(\phi^t(f^{-1}(y_0))), \tag{7.39}$$

which is Eq. (7.38).

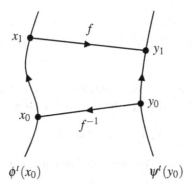

Figure 7.5 Origin of $\psi^t = f \circ \phi^t \circ f^{-1}$

7.4 LEMMA C3. LIMITING FLOW

For the diffeomorphism f of the last section, $f : M \to N$ with manifolds M and N, we use the canonical transformation S_s^{-1} from Eq. (7.1),

$$f \equiv S_s^{-1} = B_s^{-1} \circ \cdots \circ B_1^{-1} : F_0 \to F_s. \tag{7.40}$$

On F_0, the largest domain one starts with, a smooth vector field $X_0(x) \in TF_0$ shall be given, which generates a flow $\phi_0^t(x) \in F_0$. The diffeomorphism S_s^{-1} defines its push forward $X_s = (S_s^{-1})_* X_0$ on TF_s. According to the last section, this push forward generates a flow

$$\phi_s^t(x_0) = S_s^{-1} \circ \phi_0^t \circ S_s(x_0) \tag{7.41}$$

on F_s. The tangent vectors to these flows in F_0 and F_s are given by[4]

$$(X_0)_{x_0} = \left. \frac{d}{dt} \right|_{t=0} \phi_0^t(x_0), \qquad\qquad (X_s)_{x_0} = \left. \frac{d}{dt} \right|_{t=0} \phi_s^t(x_0). \tag{7.42}$$

Lemma C3 *Let F_s, B_s, S_s be as in Lemma C1. In addition to conditions (A) to (D) of Lemma C1, the following shall hold:*

   (E) *For $x \in F_\infty, |(X_s)_x - (X_\infty)_x| \leq d_{s+1}$, thus $(X_s)_x \to (X_\infty)_x$ for $s \to \infty$.*
   (F) *The segment $\bar{x} = x_0 + Yt, 0 \leq t \leq 1$ lies in F_∞ and $X_\infty = Y$ along $\bar{x}$.*
   (G) *$|\partial X_s(x)/\partial x| \leq \Theta$ for $x \in F_s$, where Θ is independent of s.*

Under the assumptions (A) to (G) one has

---

[4]Note that X_0 refers to $s = 0$, while x_0 refers to $t = 0$.

$$\phi_0^t(S_\infty(x_0)) = S_\infty(x_0 + Yt) \qquad if \qquad 0 \le t \le \frac{1}{1 + \Theta} \qquad (7.43)$$

Proof The formal correctness of Eq. (7.43) is easily seen: from (F), X_∞ generates the flow $\phi_\infty^t(x_0) = x_0 + Yt$ along the segment $x_0 + Yt$. Equation (7.43) can thus be written

$$\phi_\infty^t(x_0) = S_\infty^{-1} \circ \phi_0^t \circ S_\infty(x_0), \qquad (7.44)$$

which is Eq. (7.41) for $s \to \infty$. We show now that this limit exists. According to (E) and (F),

$$|(X_s)_{\bar{x}} - Y| \le d_{s+1} \qquad (7.45)$$

on the segment $\bar{x} = x_0 + Yt$ in F_∞ for $0 \le t \le 1$. According to assumption (B) in Lemma C1,

$$F_\infty \subset F_s - d_{s+1} - d_{s+2} - d_{s+3} - \cdots . \qquad (7.46)$$

Thus a (d_{s+1})-neighborhood of the segment $x_0 + Yt$ belongs to F_s. In this neighborhood, the distance from any point to the segment is $\le d_{s+1}$, and from the Lagrange formula (5.42) applied to (G) one has for any point x in this (d_{s+1})-neighborhood and for $\bar{x}$ on the segment,

$$|(X_s)_x - (X_s)_{\bar{x}}| \le \Theta |x - \bar{x}| \le \Theta d_{s+1}. \qquad (7.47)$$

Hence in this neighborhood, from Eqs. (7.45) and (7.47),

$$|(X_s)_x - Y| = |(X_s)_x - (X_s)_{\bar{x}} + (X_s)_{\bar{x}} - Y| \le |(X_s)_x - (X_s)_{\bar{x}}| + |(X_s)_{\bar{x}} - Y|$$
$$\le \Theta d_{s+1} + d_{s+1} = (1 + \Theta) d_{s+1}. \qquad (7.48)$$

Therefore Lemma C2 applies, with

$$\varepsilon = (1 + \Theta) d_{s+1} \qquad and \qquad d = d_{s+1}, \qquad (7.49)$$

the latter because a (d_{s+1})-neighborhood is considered. Lemma C2 gives that the solution $x(t)$ of the differential equation

$$\frac{dx(t)}{dt} = (X_s)_{x(t)}, \qquad x(0) = x_0 \qquad (7.50)$$

deviates from the straight segment $\bar{x}(t)$ starting in x_0 by

$$|x(t) - (x_0 + Yt)| \le d_{s+1} \qquad for \qquad 0 \le t \le \frac{d_{s+1}}{(1 + \Theta) d_{s+1}} = \frac{1}{1 + \Theta}. \qquad (7.51)$$

Uniqueness of the solution implies that

$$x(t) = \phi_s^t(x_0), \qquad (7.52)$$

and Eq. (7.51) becomes

$$|\phi_s^t(x_0) - (x_0 + Yt)| \le d_{s+1} \qquad \text{for} \qquad 0 \le t \le \frac{1}{1+\Theta}, \qquad (7.53)$$

which is Eq. (9) in Arnold (1963a). Since the distance between $x_0 + Yt$ and $\phi_s^t(x_0)$ is $\le d_{s+1}$, the points $\phi_s^t(x_0)$ lie in the (d_{s+1})-neighborhood of the points $\bar{x} = x_0 + Yt$. Thus the straight segment now from $x_0 + Yt$ to $\phi_s^t(x_0)$ at a given t (sufficiently small) lies in this neighborhood and, according to the above, also in F_s. The latter is the domain of B_s and S_s, and from Eq. (7.4) in Lemma C1 one has in this domain

$$\|dS_s\| \le 2^s |dx|. \qquad (7.54)$$

Hence the Lagrange formula (5.42) can be applied to S_s and the points $\phi_s^t(x_0)$ and $x_0 + Yt$, which gives with assumption (D) of Lemma C1 for $0 \le t \le (1+\Theta)^{-1}$

$$|S_s(\phi_s^t(x_0)) - S_s(x_0 + Yt)| \le 2^s |\phi_s^t(x_0) - (x_0 + Yt)| \le 2^s d_{s+1} \le \frac{2^s c}{4^{s+1}} = \frac{c}{2^{s+2}}. \qquad (7.55)$$

This means that

$$|S_s(\phi_s^t(x_0)) - S_s(x_0 + Yt)| \to 0 \quad \text{for} \quad s \to \infty \quad \text{and} \quad 0 \le t \le \frac{1}{1+\Theta}, \qquad (7.56)$$

and since $S_s(\phi_s^t(x_0)) = \phi_0^t(S_s(x_0))$ by Eq. (7.41), also

$$|\phi_0^t(S_s(x_0)) - S_s(x_0 + Yt)| \to 0 \quad \text{for} \quad s \to \infty \quad \text{and} \quad 0 \le t \le \frac{1}{1+\Theta}, \qquad (7.57)$$

as was to be shown.   □

For the material in this section, see also Wayne (1996), pp. 25-26.

7.5   LEMMA C4. MEASURE OF THE LIMIT

The last lemma of this chapter is an estimate of the measure of the limit set of the canonical transformations discussed above.

Lemma C4 *Let F be a compact region in $\mathbb{R}^n$ and (S_s) with $s = 1, 2, \dots$ a sequence of continuous maps of F onto $F_s = S_s(F) \subset R$. The maps S_s shall converge uniformly to the map S_∞ from F onto $F_\infty = S_\infty(F)$. Then*

$$\text{mes}\,(F_\infty) \ge \overline{\lim}\ \text{mes}\,(F_s). \qquad (7.58)$$

Remark Here[5]

$$\overline{\lim}\ \text{mes}\,(F_n) = \inf_{n \in \mathbb{N}} \sup_{m \ge n} \text{mes}\,(F_m) \qquad (7.59)$$

---

[5] In the following, letters m and n are used for integers instead of the iteration index s.

is the *limit superior* of the bounded sequence $(\text{mes}(F_n))$. To obtain it, one first determines the supremum of the full sequence, then that of the sequence without the first element, then that of the sequence without the first two elements, etc. The suprema of these truncated sequences converge to a unique value. Indeed, according to the Bolzano-Weierstrass theorem, every bounded sequence has a largest accumulation point, and this is exactly the limit superior. The following proof of the lemma is a standard proof from elementary measure theory, taken from Ganster (2011).

Proof Measures are *countably additive*, which means that for any sequence of pairwise disjoint sets F_n, i.e., $F_n \cap F_m = \emptyset$ for $n \neq m$,

$$\text{mes}\left(\bigcup_{n=1}^{\infty} F_n\right) = \sum_{n=1}^{\infty} \text{mes}(F_n). \tag{7.60}$$

According to the assumptions of the lemma, the sequence (F_n) converges to

$$F_\infty = \lim_{n \to \infty} F_n. \tag{7.61}$$

We prove four statements (where lim is $\lim_{n \to \infty}$):

(i) If $F_1 \subset F_2 \subset \dots$, then $\text{mes}(F_\infty) = \lim \text{mes}(F_n)$.
(ii) If $F_1 \supset F_2 \supset \dots$ and $\text{mes}(F_1) < \infty$, then $\text{mes}(F_\infty) = \lim \text{mes}(F_n)$.
(iii) $\text{mes}(F_\infty) \leq \underline{\lim} \text{mes}(F_n)$.
(iv) If $\text{mes}\left(\bigcup_{n=1}^{\infty} F_n\right) < \infty$, then $\text{mes}(F_\infty) \geq \overline{\lim} \text{mes}(F_n)$.

Statements (i) and (ii) are termed *continuity of the measure* from below and from above, respectively. Statement (iii) for the limit inferior is not needed, but included for completeness. The image $F_n = S_n(F)$ of a compact set F by a continuous map S_n is compact.[6] Thus all the F_n in the lemma have finite measure (they are bounded), and since $F_n \to F_\infty$, also $\text{mes}\left(\bigcup_{n=1}^{\infty} F_n\right) < \infty$. Therefore, statement (iv) is Lemma C4.

To show (i), for $F_1 \subset F_2 \subset \dots$ one has

$$F_\infty = \bigcup_{n=1}^{\infty} F_n. \tag{7.62}$$

Let $F_0 = \emptyset$. Then the sets $F_n \backslash F_{n-1}$ are pairwise disjoint for $n = 1, 2, \dots$, and

$$F_\infty = \bigcup_{n=1}^{\infty} F_n \backslash F_{n-1}. \tag{7.63}$$

---

[6]Proof: consider a continuous map $f : X \to Y$ with metric spaces X, Y. Let $A \subset X$ be compact and B_i, $i \in I$ with index set I be an open covering of B. Then $f^{-1}(B_i)$ is an open covering of A. Since A is compact, already $A \subset \bigcup_{j \in J} f^{-1}(B_j)$ for some finite subset $J \subset I$. Then

$$B = f(A) \subset f\left(\bigcup_{j \in J} f^{-1}(B_j)\right) = \bigcup_{j \in J} f(f^{-1}(B_j)) = \bigcup_{j \in J} (B_j \cap f(X)) \subset \bigcup_{j \in J} B_j.$$

Thus the finite family B_j with $j \in J$ is an open covering of B, hence B is compact.

Therefore, using Eq. (7.60),

$$\text{mes}\,(F_\infty) = \text{mes}\left(\bigcup_{n=1}^{\infty} F_n \backslash F_{n-1}\right)$$

$$= \sum_{n=1}^{\infty} \text{mes}\,(F_n \backslash F_{n-1})$$

$$= \lim_{m\to\infty} \sum_{n=1}^{m} \text{mes}\,(F_n \backslash F_{n-1})$$

$$= \lim_{m\to\infty} \text{mes}\left(\bigcup_{n=1}^{m} F_n \backslash F_{n-1}\right)$$

$$= \lim_{m\to\infty} \text{mes}\,(F_m). \qquad (7.64)$$

(ii) Here $F_1 \supset F_2 \supset \ldots$, and

$$F_\infty = \bigcap_{n=1}^{\infty} F_n. \qquad (7.65)$$

Put $E_n = F_1 \backslash F_n$ for $n = 1,2,\ldots$. Then $E_1 \subset E_2 \subset \ldots$ with $E_n \to E_\infty = F_1 \backslash F_\infty$, and (i) is applicable to the sequence (E_n). Using $F_n = F_1 \backslash (F_1 \backslash F_n) = F_1 \backslash E_n$ and the de Morgan rule (with arbitrary index set $\{n\}$)

$$\overline{\bigcap F_n} = \bigcup \overline{F_n} \qquad (7.66)$$

with $\overline{F_n} = F_1 \backslash F_n = E_n$, one has

$$\bigcap F_n = \overline{\bigcup E_n} = F_1 \backslash \bigcup E_n. \qquad (7.67)$$

Therefore,

$$\text{mes}\,(F_\infty) = \text{mes}\left(\bigcap_{n=1}^{\infty} F_n\right)$$

$$= \text{mes}\left(F_1 \backslash \bigcup_{n=1}^{\infty} E_n\right)$$

$$\overset{(7.60)}{=} \text{mes}\,(F_1) - \text{mes}\left(\bigcup_{n=1}^{\infty} E_n\right)$$

$$= \text{mes}\,(F_1) - \text{mes}\,(E_\infty)$$

$$\overset{(7.64)}{=} \text{mes}\,(F_1) - \lim_{n\to\infty} \text{mes}\,(E_n)$$

$$= \lim_{n\to\infty} [\text{mes}\,(F_1) - \text{mes}\,(E_n)]$$

$$\overset{(7.60)}{=} \lim_{n\to\infty} \text{mes}\,(F_1 \backslash E_n)$$

$$= \lim_{n\to\infty} \text{mes}\,(F_n). \qquad (7.68)$$

To avoid in $\mathrm{mes}\,(F_1) - \mathrm{mes}\,(E_\infty)$ an unallowed $\infty - \infty$, one has to demand that $\mathrm{mes}\,(F_1) < \infty$.

(iii) Let $E_n = \bigcap\limits_{k=n}^{\infty} F_k$. Then $E_1 \subset E_2 \subset \ldots$ and $\lim F_n = \lim E_n$, which gives

$$\mathrm{mes}\,(F_\infty) \overset{(a)}{=} \lim\,\mathrm{mes}\,(E_n) \overset{(b)}{=} \underline{\lim}\,\mathrm{mes}\,(E_n) \overset{(c)}{\leq} \underline{\lim}\,\mathrm{mes}\,(F_n), \qquad (7.69)$$

where (a) is Eq. (7.64), (b) uses $\lim a_n = \underline{\lim} a_n$ for any convergent sequence (a_n) with $a_n \in \mathbb{R}$ (here $a_n \equiv \mathrm{mes}\,(E_n)$) and (c) uses that $E_n \subset F_n$ implies $\mathrm{mes}\,(E_n) \leq \mathrm{mes}\,(F_n)$.

(iv) Let $E_n = \bigcup\limits_{k=n}^{\infty} F_k$. Then $E_1 \supset E_2 \supset \ldots$ and $\lim F_n = \lim E_n$, which gives

$$\mathrm{mes}\,(F_\infty) \overset{(a)}{=} \lim\,\mathrm{mes}\,(E_n) \overset{(b)}{=} \overline{\lim}\,\mathrm{mes}\,(E_n) \overset{(c)}{\geq} \overline{\lim}\,\mathrm{mes}\,(F_n), \qquad (7.70)$$

where (a) uses Eq. (7.68), which is allowed since $\mathrm{mes}\,(E_1) < \infty$, (b) uses $\lim a_n = \overline{\lim} a_n$ for any convergent sequence (a_n) and (c) uses that $E_n \supset F_n$ implies $\mathrm{mes}\,(E_n) \geq \mathrm{mes}\,(F_n)$. □

8 Arithmetic Lemmas

8.1 LEMMA D1. MEASURE OF BOUNDARY LAYER

The following seven arithmetic lemmas D1 to D7 (the 'D' refers to the surface-to-volume ratio that appears in some of them) can be found in section 4.1 of Arnold (1963a). This is a central section, but Arnold is quite brief there. He derives in these lemmas the measure of the frequency domain that remains after the removal of boundary layers and interior strips from a domain Ω. These removals are required for estimates of Fourier coefficients from complex function theory and in order to avoid resonances.

Let Ω be a simply connected, bounded and closed domain of dimension n in $\mathbb{R}^n$ with Cartesian coordinates $\omega_1, \ldots, \omega_n$ for points $\omega \in \Omega$. Its boundary $\partial\Omega$ shall 'consist of a finite number of pieces of smooth manifolds' (Arnold 1963a, p. 24). Let $\Omega - d$ be the domain Ω with a layer of width d removed from its boundary as defined in Eq. (2.19).

Lemma D1 *An estimate for the upper limit of the volume of this layer is given by*

$$\boxed{\text{mes}\,(\Omega \backslash (\Omega - d)) \leq D\,\text{mes}\,(\Omega)\,d} \tag{8.1}$$

Proof

$$
\begin{aligned}
\text{mes}\,(\Omega \backslash (\Omega - d)) &= \int_0^d dl\,\text{mes}\,(\partial(\Omega - l)) \\
&\overset{(\P)}{\leq} \text{mes}\,(\partial\Omega) \int_0^d dl \\
&= \frac{\text{mes}\,(\partial\Omega)}{\text{mes}\,(\Omega)}\,\text{mes}\,(\Omega)\,d \\
&= D\,\text{mes}\,(\Omega)\,d.
\end{aligned}
\tag{8.2}
$$

The estimate ($\P$) holds also if Ω is not convex but has concave indentations, since the closed surface $\partial(\Omega - l)$ inside $\partial\Omega$ has smaller measure than $\partial\Omega$. $\qquad\square$

Remark 1 If $\partial\Omega$ is strongly folded, i.e., Ω has many alternating convex and concave portions, then mes $(\partial\Omega)$ and thus D can become arbitrarily large. In the proof of the KAM theorem, the number D appears in measure estimates only (see Eqs. 4.152, 4.157 and 4.161) and in the bound of the variable δ (see Eqs. 4.107 and 4.197).[1]

---

[1] The surface-to-volume ratio plays an important role in biological cells. A large surface-to-volume ratio (in small cells) makes *diffusion* more efficient; this may be the reason for Arnold's choice of the letter D.

DOI: 10.1201/9781003287803-8

Remark 2 With appropriate changes, the estimate (8.1) also applies to multiply connected domains Ω. Consider for example a doubly connected domain $\Omega = \Omega' \backslash T$ with simply connected Ω' and 'tunnel' T. Choose $r > 1$ so that

$$\Omega \subset rT, \tag{8.3}$$

where

$$\omega \in rT \quad \leftrightarrow \quad \omega/r \in T, \tag{8.4}$$

with ω understood as Cartesian n-tuple and $\omega = 0$ inside T. Then

$$\begin{aligned} \mathrm{mes}\,(\Omega \backslash (\Omega - d)) &\le \big(\mathrm{mes}\,(\partial\Omega') + r\,\mathrm{mes}\,(\partial T)\big)\,d \\ &= \frac{\mathrm{mes}\,(\partial\Omega') + r\,\mathrm{mes}\,(\partial T)}{\mathrm{mes}\,(\Omega)}\,\mathrm{mes}\,(\Omega)\,d \\ &= D\,\mathrm{mes}\,(\Omega)\,d. \end{aligned} \tag{8.5}$$

This is easily generalized to n-fold connected domains. In the following, only simply connected domains are considered.

8.2 LEMMA D2. MEASURE OF INNER LAYER

The next lemma is a straightforward estimate.

Lemma D2 *Let $d_2 > d_1$. Then*

$$\boxed{\mathrm{mes}\,((\Omega - d_1)\backslash(\Omega - d_2)) \le D\,\mathrm{mes}\,(\Omega)\,(d_2 - d_1)} \tag{8.6}$$

Proof The following estimate holds, see Fig. 8.1,

$$\mathrm{mes}\,((\Omega - d_1)\backslash(\Omega - d_2)) \le \mathrm{mes}\,(\Omega \backslash (\Omega - (d_2 - d_1))) \tag{8.7}$$

$$(\Omega - d_1)\backslash(\Omega - d_2) \qquad\qquad \Omega \backslash (\Omega - (d_2 - d_1))$$

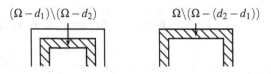

Figure 8.1   See Eq. (8.7), here for $d_2 = 2d_1$

As noted above, this is also true if Ω is not convex. Putting $d = d_2 - d_1$ in Eq. (8.1) and using Eq. (8.7) one obtains Eq. (8.6).   □

8.3   LEMMA D3. MEASURE OF STRIP

A *hyperplane* in the space $\mathbb{R}^n$ of frequency vectors ω is given by the linear relation

$$\sum_{i=1}^{n} k_i \omega_i = 0. \tag{8.8}$$

In the following, only hyperplanes with coefficients $k_i \in \mathbb{Z}$ are considered, and Eq. (8.8) becomes a resonance condition for the ω_i. As observed before, resonances can cause the infinite sequence of canonical transformations of the perturbed Hamiltonian to diverge. All resonances will be avoided by removing planar *strips* from the frequency domain Ω, i.e., neighborhoods of hyperplanes of constant width. Since the max norm $|\omega| = \max_i |\omega_i|$ is used, a few unorthodox formulas result for distances and volumes in ω-space.

For a given k-vector with $|k| = |k_1| + \cdots + |k_n| \geq 1$, the inequality $|(k, \omega)| < h/2$ defines a strip or slab of total width $< h$ in ω-space (both sides of the hyperplane $|(k, \omega)| = 0$). For example, if $k = (k_i) = (\delta_{ij})$ for a fixed j, the strip is given by $|\omega_j| < h/2$ or $\omega_j \in (-h/2, h/2)$ and has width h. However, if $k = (1, -1, 0, \ldots, 0)$, the strip is given by $|\omega_1 - \omega_2| < h/2$, i.e., is an $(h/2)$-neighborhood of the line $\omega_2 = \omega_1$. Its total width is again h in the max-norm, but $\sqrt{2}h$ in the Pythagorean 2-norm.[2]

Lemma D3 *Let Ω be a simply connected, bounded and closed domain in $\mathbb{R}^n$ and Γ be a strip of width h. Then*

$$\boxed{\operatorname{mes}(\Omega \cap \Gamma) \leq D \operatorname{mes}(\Omega)\, nh} \tag{8.9}$$

Remark The factor nh is for the max norm. The (expected) factor $\sqrt{n}h$ holds for the 2-norm.

Proof (i) Consider first the case $n = 3$ (Fig. 8.2), where the space diagonal of the coordinate cube $\hat{\omega}_1, \hat{\omega}_2, \hat{\omega}_3$ is normal to the planes $\partial \Gamma$ (angles are defined using the inner product of an Euclidean space). The distance between the points $P = (0,0,0)$ and $Q = (h, h, h)$ is $\operatorname{dist}(P, Q) = |(0,0,0) - (h,h,h)| = \max(h,h,h) = h$. The distance l between points P and R on the ω_1-axis is calculated from the projection (2-norm) on the cube diagonal,

$$(l,0,0) \cdot \frac{1}{\sqrt{3}} \begin{pmatrix} 1 \\ 1 \\ 1 \end{pmatrix} = \sqrt{3}h, \tag{8.10}$$

giving $l = 3h$. All three coordinate axes $\omega_1, \omega_2, \omega_3$ extend over this distance $3h$ inside the strip of width h. For arbitrary n one obtains $l = nh$. If the cube diagonal is *not*

---

[2] In vector spaces of finite dimension, the max norm $\|.\|_\infty$ and the 2-norm $\|.\|_2$ are *equivalent*, since for arbitrary x one has $\|x\|_\infty \leq \|x\|_2 \leq \sqrt{n}\,\|x\|_\infty$.

normal to $\partial\Gamma$, there is clearly at least one direction ω_j – choose it to be ω_1 – that crosses the h-strip over a distance[3] $\leq nh$.

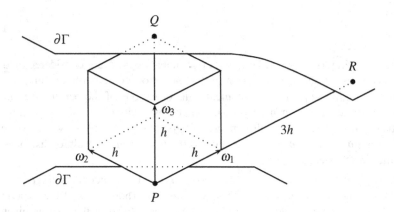

Figure 8.2 Cube of side length h along coordinate axes $\omega_1, \omega_2, \omega_3$. The space diagonal $\overline{PQ}$ of the cube is normal to $\partial\Gamma$ and has length h in the max norm. $\hat{\omega}_1, \hat{\omega}_2, \hat{\omega}_3$ cut $\partial\Gamma$ at $\overline{PR} = 3h$

(ii) Let $A = \{(\omega_2, \ldots, \omega_n)\}$ be an $(n-1)$-dimensional hyperspace. As above, the axis ω_1 shall extent over a length $\leq nh$ inside Γ. Then (Fig. 8.3)

$$\mathrm{mes}\,(\Omega \cap \Gamma) = \int_A \mathrm{d}a\, l \leq nh \int_{\partial\Omega} \mathrm{d}a' = \mathrm{mes}\,(\partial\Omega)\, nh = D\,\mathrm{mes}\,(\Omega)\, nh. \qquad (8.11)$$

As usual in the integration over non-convex boundaries (with folds or indentations), multiplicities in the substitution from $\mathrm{d}a$ in A to $\mathrm{d}a'$ on $\partial\Omega$ pose no problem, if counted with the appropriate sign; even without accounting for signs, Eq. (8.11) remains correct as an upper bound. □

8.4 LEMMA D4. MEASURE OF LAYERS AND STRIPS

Lemma D4 *Let Ω' be a domain obtained by first removing a boundary layer of width d and then N strips of maximum width h from a domain Ω,*

$$\Omega' = (\Omega - d) \backslash \bigcup_{i=1}^{N} \Gamma_i. \qquad (8.12)$$

Let $d_2 > d_1 > 0$. Then

---

[3] 'It is easy to calculate that the intersection of a strip of width h with one of the n coordinate axes of ω is not longer than nh.' (Arnold 1963b, p. 164)

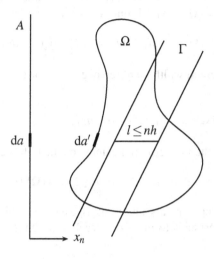

Figure 8.3   Removing a strip Γ of width h from a domain Ω

$$\text{mes}\left((\Omega' - d_1)\backslash(\Omega' - d_2)\right) \le D(1 + 2nN)\,\text{mes}\,(\Omega)\,(d_2 - d_1) \tag{8.13}$$

Remark The order in Eq. (8.12) is significant: first a boundary layer is removed, then N resonance strips. Equation (8.13) addresses the next boundary layer removal in the sequence of canonical transformations. See also Lemma D7 below.

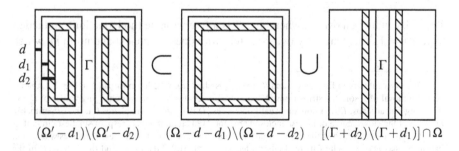

Figure 8.4   Removing boundary layers of width d, d_1, d_2 and a planar strip Γ from Ω. The subset relation $\subset$, the set union $\cup$ and the expressions under the panels refer to the hatched regions

Proof Consider first the case that a boundary layer and then a single strip Γ is removed from Ω, i.e., $\Omega' = (\Omega - d)\backslash\Gamma$. From Fig. 8.4, one obtains

$$(\Omega' - d_1)\backslash(\Omega' - d_2) \subset (\Omega - d - d_1)\backslash(\Omega - d - d_2) \cup ([(\Gamma + d_2)\backslash(\Gamma + d_1)] \cap \Omega). \tag{8.14}$$

If then N strips of maximum width h are removed, i.e., for Ω' as in Eq. (8.12), one has instead[4]

$$(\Omega' - d_1)\backslash(\Omega' - d_2) \subset (\Omega - d - d_1)\backslash(\Omega - d - d_2) \cup ([(\cup\Gamma_i + d_2)\backslash(\cup\Gamma_i + d_1)] \cap \Omega). \tag{8.15}$$

Equation (8.6) gives for the first term in the union on the right side

$$\text{mes}((\Omega - d - d_1)\backslash(\Omega - d - d_2)) \leq D\,\text{mes}(\Omega)\,(d_2 - d_1). \tag{8.16}$$

Regarding the second term in this union, as is clear from Fig. 8.4, each $[(\Gamma_i + d_2)\backslash(\Gamma_i + d_1)] \cap \Omega$ corresponds to *two* strips, each of width $d_2 - d_1$, for which Eq. (8.11) gives

$$\text{mes}([(\Gamma_i + d_2)\backslash(\Gamma_i + d_1)] \cap \Omega) \leq D\,\text{mes}(\Omega)\,2n(d_2 - d_1). \tag{8.17}$$

Thus for N (partially overlapping) layers,

$$\text{mes}([(\cup\Gamma_i + d_2)\backslash(\cup\Gamma_i + d_1)] \cap \Omega) \leq D\,\text{mes}(\Omega)\,2nN(d_2 - d_1). \tag{8.18}$$

Adding Eqs. (8.16) and (8.18) gives Eq. (8.13).[5]   □

8.5   LEMMA D5. INTEGRAL VECTORS WITH GIVEN NORM

Lemma D5 *The number of vectors $k \in \mathbb{Z}^n$ with $|k| = |k_1| + \cdots + |k_n| = m \geq 1$ is*

$$\leq 2^n m^{n-1}. \tag{8.19}$$

Proof By induction over n, where $m \geq 1$ is a free variable (no induction). Induction start is $n = 1$. There are only two k-vectors with one component and norm m, which

---

[4]There is a subtlety in Eqs. (8.14) and (8.15). As is clear from the leftmost panel in Fig. 8.4, the repeated removal of resonance strips from Ω leaves frequency domains of ever changing values of D. By removing planar slabs, D of a domain usually increases: D is minimal for spheres (cf. the isoperimetric problem of the closed curve of largest enclosed area for fixed circumference, resp. of smallest circumference for given area, which is the circle); thus, when a ball is cut into two halves, D increases (the unit ball in $\mathbb{R}^3$ has $D = 3$, each of its half-spheres has $D = 4.5$). Still, Eqs. (8.14) and (8.15) have D of the initial frequency domain Ω during all subsequent removals: the middle and right panel in Fig. 8.4 (and the corresponding terms in the equations) relate the removed layer and strip measures to this full, initial domain Ω.

[5]This is Eq. (3) in Arnold (1963a). The misprint hN there should be replaced by $2nN$, see also Eq. (5.1.3) in Arnold (1963b).

are $k = (m)$ and $k = (-m)$. Thus Eq. (8.19) is correct for $n = 1$, with $2 \leq 2^1 \cdot 1 = 2^n m^{n-1}$.

For $n = 2$, the number of vectors (k_1, k_2) with $|k| = m$ is $4m$: there are $m - 1$ vectors $k = (1, m-1), (2, m-2), \ldots, (m-1, 1) \in \mathbb{N}^2$ of norm m. If both components can be positive or negative, this gives $4(m-1)$ vectors. Furthermore, there are four vectors $k = (0, m), (0, -m), (m, 0)$ and $(-m, 0)$ of norm m, giving a total of $4(m-1) + 4 = 4m$ vectors. Equation (8.19) is again correct, since $4m \leq 2^2 \cdot m^1 = 2^n m^{n-1}$. This result will be used in the critical case $n = m = 2$ encountered below.

The induction step is from n to $n+1$ and, first, for $m \geq 2$; the case $m = 1$ is treated separately at the end of the proof. Let all $k_i \geq 0$ and assume as induction hypothesis that for $m \geq 2$ there are

$$\leq m^{n-1} \tag{8.20}$$

vectors with $k_1 + \cdots + k_n = m$. We show that Eq. (8.20) holds for $n+1$ and $m \geq 2$, i.e., the number of vectors with $k_1 + \cdots + k_n + k_{n+1} = m$ is $\leq m^n$.

The new component k_{n+1} can have any value from 0 to m. If its value is 0, then the first n components of k can still have all their $\leq m^{n-1}$ combinations adding up to $|k| = m$ from the induction hypothesis, since $k_{n+1} = 0$ does not change $|k|$. If $k_{n+1} = 1$, then for k_1 to k_n all combinations are allowed with $k_1 + \cdots + k_n = m - 1$. According to the induction hypothesis, which is true for all $m \geq 2$, there are $\leq (m-1)^{n-1}$ such combinations. Similarly, if $k_{n+1} = 2$, then k_1 to k_n give $\leq (m-2)^{n-1}$ combinations with $|k_1| + \cdots + |k_n| = m - 2$, etc., so that the total number of vectors with $k_1 + \cdots + k_{n+1} = m$ is

$$\leq m^{n-1} + (m-1)^{n-1} + (m-2)^{n-1} + \cdots + 2^{n-1} + n + 1, \tag{8.21}$$

where, from left to right, the corresponding last component is $k_{n+1} = 0, 1, 2, \ldots, m-2, m-1, m$. The final '1' in Eq. (8.21) is the number of positive vectors with $|k| = m$ and $k_{n+1} = m$, which is one, namely $k = (0, \ldots, 0, m)$. Similarly, the term n near the end of Eq. (8.21) is the number of vectors with $|k| = m$ and $k_{n+1} = m - 1$, which is n, namely $k = (1, 0, \ldots, 0, m-1)$ to $k = (0, \ldots, 0, 1, m-1)$. For both these cases, Eq. (8.20) does *not* apply: for $k_1 + \cdots + k_n = 0$ it gives the wrong number 0 instead of 1 possible vector, $(0, \ldots, 0)$, and for $k_1 + \cdots + k_n = 1$ it gives a wrong number 1 instead of n (this failure is avoided in Eq. 8.19 because of the factor 2^n allowing for a sign change of the components). We must therefore assume $m \geq 2$ in Eq. (8.20).

The sum in Eq. (8.21) has $m + 1$ terms. For the last two, one has

$$n + 1 \leq m^{n-1}, \tag{8.22}$$

except for $n = 1, m \geq 1$, which was treated above as induction start, and except for the single case $n = m = 2$, which was addressed in the subsequent paragraph for $n = 2, m \geq 1$. In all other cases, Eq. (8.21) has m terms when one counts the final $n + 1$ as one term, each of which is $\leq m^{n-1}$, thus this sum is $\leq m^n$, as was to be shown.

Allowing now for the k_i to have negative signs, the number of possible vectors $(k_1, \ldots, k_n)$ with $|k| = m \geq 2$ is

$$\leq 2^n m^{n-1}, \tag{8.23}$$

since there are at most 2^n sign combinations (there are less if $k_i = 0$ for some i).

Finally, we address the case $m = 1$ for which Eq. (8.20) does not apply. There are n vectors $k = (1, 0, \ldots, 0)$ to $(0, \ldots, 0, 1)$ in $\mathbb{N}^n$ with $|k| = 1$. Allowing for a sign change, there are $2n$ vectors, showing that Eq. (8.19), which reads then here $2n \leq 2^n$, is also correct for $n \geq 1$ and $m = 1$. Thus, all cases are covered and the lemma is proved.   □

8.6   LEMMA D6. INTEGRAL VECTORS BELOW GIVEN NORM

Lemma D6 *The number of vectors $k \in \mathbb{Z}^n$ with* [6] $1 \leq |k| \leq m$ *where $m \geq 1$ is*

$$\leq 2^n m^n. \tag{8.24}$$

Proof According to Lemma D5, this number of vectors has as upper bound the sum of the m terms $2^n p^{n-1}$ for $1 \leq p \leq m$, i.e.,

$$2^n + 2^n 2^{n-1} + 2^n 3^{n-1} + \cdots + 2^n m^{n-1}. \tag{8.25}$$

Each term in this sum is $\leq 2^n m^{n-1}$, thus the sum has the upper bound

$$m \, 2^n m^{n-1} = 2^n m^n, \tag{8.26}$$

as was to be shown.   □

8.7   LEMMA D7. MEASURE OF LOST DOMAIN

The planar strips removed from the frequency domain Ω correspond to resonances with $|(k, \omega)| < h/2$. For $|k| \to \infty$, an infinite number of such strips is removed. The domain Ω of type D is bounded. In order that only a finite measure is removed from Ω by an infinite number of resonances, the strip width h must decrease with increasing $|k|$. This is achieved by a *Diophantine condition*,

$$\boxed{|(k, \omega)| \geq \frac{K}{|k|^{n+1}}} \tag{8.27}$$

which defines the domain outside resonances. Here $K > 0$ is a constant and $n > 1$ the number of degrees of freedom. One says that a vector ω that obeys Eq. (8.27) is

---

[6] Arnold (1963a,b) states only $|k| \leq m$, but $|k| = 0$ has to be excluded: including it, Eq. (8.24) is wrong for $n = 1$ and all m, and for $(n, m) = (2, 1)$.

of type $(K, n+1)$, see Wayne (1996), pp. 6, 16. The resonance domain defined by $|(k, \omega)| < K/|k|^{n+1}$ is a strip of width

$$h = 2K/|k|^{n+1}. \tag{8.28}$$

Let $N > 1$ be a natural number and define

$$\Omega_{KN} = \{\omega \in \Omega : |(k, \omega)| \geq K/|k|^{n+1} \text{ for } 0 < |k| < N\}. \tag{8.29}$$

Consider infinite sequences of numbers $1 < N_1 < N_2 < \ldots$ and $d_1, d_2, \ldots > 0$. For $s = 1, 2, \ldots$ let

$$\Omega_s = (\Omega_{s-1})_{KN_s} - d_s, \qquad \Omega_0 = \Omega. \tag{8.30}$$

Thus for $s \geq 1$, all resonance strips with $|k| < N_s$ are removed. Thereafter, a layer of width d_s is removed from the boundary of the remaining set, including the boundaries of all strips removed so far.

Lemma D7 *For* $s = 1, 2, \ldots$ *and* $d_s > 0$ *one has, with* Ω_s *defined in Eq.* (8.30),

$$\boxed{\operatorname{mes}(\Omega_{s-1} \backslash \Omega_s) \leq LD(K\sigma_s + N_s^n d_s) \operatorname{mes}(\Omega)} \tag{8.31}$$

where

$$\sigma_s = \sum_{m=N_{s-1}}^{N_s-1} \frac{1}{m^2}, \qquad L = n2^{n+2}, \tag{8.32}$$

and $N_0 = 1$.

Proof For arbitrary sets $C \subset B \subset A$, one has

$$A \backslash C = (A \backslash B) \cup (B \backslash C). \tag{8.33}$$

Indeed,

$$
\begin{aligned}
A \backslash C &= ((A \backslash B) \cup B) \backslash C && \text{since } B \subset A \\
&= (A \backslash B) \backslash C \cup B \backslash C && \text{since } (A \backslash B) \cap B = \emptyset \\
&= (A \backslash B) \cup (B \backslash C) && \text{since } C \subset B.
\end{aligned} \tag{8.34}
$$

The middle equation follows from the distributive law for $\wedge$ and $\vee$.[7] Since $(\Omega_{s-1})_{KN_s} - d_s \subset (\Omega_{s-1})_{KN_s} \subset \Omega_{s-1}$, one has thus

$$\Omega_{s-1} \backslash [(\Omega_{s-1})_{KN_s} - d_s]$$
$$= (\Omega_{s-1} \backslash (\Omega_{s-1})_{KN_s}) \cup ((\Omega_{s-1})_{KN_s} \backslash [(\Omega_{s-1})_{KN_s} - d_s]), \tag{8.35}$$

---

[7] For arbitrary statements P, Q, R one has $(P \vee Q) \wedge R \leftrightarrow (P \wedge R) \vee (Q \wedge R)$. Thus for all sets A', B, C (here $A' = A \backslash B$) one has $(A' \cup B) \backslash C = \{x : (x \in A' \vee x \in B) \wedge x \notin C\} = \{x : (x \in A' \wedge x \notin C) \vee (x \in B \wedge x \notin C)\} = (A' \backslash C) \cup (B \backslash C)$.

and therefore, with $\Omega_s = (\Omega_{s-1})_{KN_s} - d_s$,

$$\text{mes}\,(\Omega_{s-1}\backslash\Omega_s)$$
$$\leq \text{mes}\,(\Omega_{s-1}\backslash(\Omega_{s-1})_{KN_s}) + \text{mes}\,((\Omega_{s-1})_{KN_s}\backslash[(\Omega_{s-1})_{KN_s} - d_s]). \quad (8.36)$$

Here

$$\text{mes}\,(\Omega_{s-1}\backslash(\Omega_{s-1})_{KN_s}) \leq \sum_{k=N_{s-1}}^{N_s-1} \text{mes}\,(\Omega\cap\Gamma_k), \quad (8.37)$$

where the sum on the right side is the measure of all strips with $N_{s-1} \leq |k| < N_s$ removed from Ω. From Eq. (8.28), the width of the strip Γ_k corresponding to the resonance $k = (k_1,\ldots,k_n)$ is $h = 2K/|k|^{n+1}$; and according to Eq. (8.19), for $|k| = m$ there are at most $2^n m^{n-1}$ such vectors k and thus strips Γ_k. Their total measure according to Eq. (8.9) is, setting $|k| = m$ in Eq. (8.28),

$$\sum_{|k|=m} \text{mes}\,(\Omega\cap\Gamma_k) \leq 2^n m^{n-1}\, D\, \text{mes}\,(\Omega)\, n\, \frac{2K}{m^{n+1}}$$
$$= \frac{n2^{n+1}}{m^2}\, DK\, \text{mes}\,(\Omega) < \frac{L}{m^2}\, DK\, \text{mes}\,(\Omega), \quad (8.38)$$

with $L = n2^{n+2}$. Summation over m in the bounds $N_{s-1} \leq m < N_s$ gives in Eq. (8.37)

$$\text{mes}\,(\Omega_{s-1}\backslash(\Omega_{s-1})_{KN_s}) \leq LDK \left(\sum_{m=N_{s-1}}^{N_s-1} \frac{1}{m^2}\right) \text{mes}\,(\Omega) = LDK\sigma_s\, \text{mes}\,(\Omega). \quad (8.39)$$

Note that, 'as $K \to 0$, the total measure of resonance zones tends to zero' (Arnold 1963b, p. 98).

The second term in the sum on the right side of Eq. (8.36) is obtained from Eq. (8.13) by putting $d_2 = d_s$ and $d_1 = 0$. In Eq. (8.13), N is the total number of strips removed from Ω so far, which is the number of vectors k with $|k| < N_s$. From Eq. (8.24), the number of vectors k with $|k| \leq N_s$ is at most $2^n N_s^n$. Therefore,

$$\text{mes}\,((\Omega_{s-1})_{KN_s}\backslash[(\Omega_{s-1})_{KN_s} - d_s]) \leq D\,(1 + 2n2^n N_s^n)\,d_s\, \text{mes}\,(\Omega)$$
$$< n2^{n+2}\, DN_s^n d_s\, \text{mes}\,(\Omega)$$
$$= LDN_s^n d_s\, \text{mes}\,(\Omega). \quad (8.40)$$

Adding Eqs. (8.39) and (8.40) gives Eq. (8.31). $\qquad\square$

References

Abraham, R. and J. E. Marsden 1978, *Foundations of mechanics*, 2nd edition, Reading, Massachusetts, Benjamin/Cummings Publishing Company

Ahlfors, L. V. 1966, *Complex analysis*, New York, McGraw-Hill

Aleksandrov, P. S. 1956, *Combinatorial topology*, vol. 1, Rochester, N. Y., Graylock Press

Anosov, D. V. 1995, *Differential equations on a torus*, in *Encyclopaedia of Mathematics*, vol. 2, ed. M. Hazewinkel, Dordrecht, Springer, pp. 217-219

Arnold, V. I. 1961, *Small denominators. I. Mappings of the circumference onto itself*, Izvestiya Rossiiskoi Akademii Nauk SSSR Ser. Mat., 25, pp. 21-86 (Russian), American Mathematical Society Translations (2) 46 (1965), pp. 213-284

Arnold, V. I. 1963a, *Proof of a theorem of A. N. Kolmogorov on the invariance of quasi-periodic motions under small perturbations of the Hamiltonian*, Uspekhi Mat. Nauk, 18, pp. 13-40 (Russian), Russian Mathematical Surveys, 18, pp. 9-36 (English)

Arnold, V. I. 1963b, *Small denominators and problems of stability of motion in classical and celestial mechanics*, Uspekhi Mat. Nauk, 18, pp. 91-192 (Russian), Russian Mathematical Surveys, 18, pp. 85-191 (English)

Arnold, V. I. 1963c, *On a theorem of Liouville concerning integrable problems of dynamics*, Sibirskii Matematicheskii Zhurnal, 4:2, English translation published in American Mathematical Society Translations (2) 61 (1967), pp. 292-296

Arnold, V. I. 1965, *Stability and instability in classical mechanics*, Second Math. Summer School, Part II (in Russian), Kiev, Naukova Dumka, pp. 85-119. Reprinted in *V. I. Arnold, Collected Works*, vol. 1, Berlin, Springer

Arnold, V. I. 1972, *Book Review: Celestial mechanics I, II by Shlomo Sternberg*, Bulletin of the American Mathematical Society, 78(6), pp. 962-963

Arnold, V. I. 1989, *Mathematical methods of classical mechanics*, New York, Springer

Arnold, V. I. and A. Avez 1968, *Ergodic problems of classical mechanics*, New York, W. A. Benjamin

Arnold, V. I., V. V. Kozlov and A. I. Neishtadt 1997, *Mathematical aspects of classical and celestial mechanics*, Berlin, Springer

Arrowsmith, D. K. and C. M. Place 1990, *An introduction to dynamical systems*, Cambridge, Cambridge University Press

Borsuk, K. 1933, *Drei Sätze über die n-dimensionale euklidische Sphäre*, Fundamenta Mathematicae, 20, pp. 177-190

Bott, R. and L. Tu 1982, *Differential forms in algebraic topology*, New York, Springer

Brent, R. P., J. H. Osborn and W. D. Smith 2014, *Bounds on determinants of perturbed diagonal matrices*, arXiv:1401.7084v7, 18 pp.

Broer, H. W. and M. B. Sevryuk 2010, *KAM theory: quasi-periodicity in dynamical systems*, in *Handbook of dynamical systems*, vol. 3, ed. H. W. Broer, B. Hasselblatt, F. Takens, Amsterdam, Elsevier, pp. 249-344

Brouwer, L. E. J. 1911a, *Beweis der Invarianz der Dimensionenzahl*, Mathematische Annalen, 70, pp. 161-165

Brouwer, L. E. J. 1911b, *Über Abbildung von Mannigfaltigkeiten*, Mathematische Annalen, 71, pp. 97-114

Brouwer, L. E. J. 1976, *Collected works*, vol. 2, *Geometry, analysis, topology and mechanics*, ed. Hans Freudenthal, Amsterdam, North-Holland

Cantor, G. 1877, *Ein Beitrag zur Mannigfaltigkeitslehre*, Crelles Journal für die reine und angewandte Mathematik, 84, pp. 242-258

Chierchia, L. 2009, *Kolmogorov–Arnold–Moser (KAM) theory*, in *Encyclopedia of complexity and systems science*, ed. R. A. Meyers, New York, Springer, https://doi.org/10.1007/978-0-387-30440-3_302

Ćuk, M., D. P. Hamilton and M. J. Holman 2012, *Long-term stability of horseshoe orbits*, Monthly Notices of the Royal Astronomical Society, 426, pp. 3051-3056

de la Llave, R. 2001, *A tutorial on KAM Theory*, in *Proceedings of Symposia in Pure Mathematics*, vol. 69, Summer school on *Smooth ergodic theory and its applications, Seattle, July-August 1999*, ed. A. Katok, R. de la Llave, Ya. Pesin & H. Weiss, Providence, R.I., American Mathematical Society, pp. 175-292, pdf file available at different locations [this version cited here]

Deift, P. A. 2019, *Three lectures on "Fifty years of KdV: An integrable system,"* in *Nonlinear dispersive partial differential equations and inverse scattering*, ed. P. D. Miller, P. Perry, J.-C. Saut and C. Sulem, Fields Institute Communications 83, New York, Springer, pp. 3-38

Dieudonné, J. 1969, *Foundations of modern analysis*, New York, Academic Press

Dieudonné, J. 1976, *Grundzüge der modernen Analysis*, Band 3, Braunschweig, Vieweg

Dieudonné, J. 1989, *A history of algebraic and differential topology, 1900-1960*, Boston, Birkhäuser

Dolzhenko, E. P. 1995, *Monogenic function*, in *Encyclopaedia of Mathematics*, vol. 4, ed. M. Hazewinkel, Dordrecht, Springer, pp. 7-8

Dugundji, J. and A. Granas 1982, *Fixed point theory*, vol. 1, Warszawa, Polish Scientific Publishers

Dumas, H. S. 2014, *The KAM story: A friendly introduction to the content, history, and significance of classical Kolmogorov–Arnold–Moser theory*, Singapore, World Scientific

Eliasson, L. H. 1996, *Absolutely convergent series expansions for quasi periodic motions*, Mathematical Physics Electronic Journal, 2, paper 4, pp. 1-33 (electronic)

Faltings, G. 1983, *Endlichkeitssätze für abelsche Varietäten über Zahlkörpern*, Inventiones mathematicae, 73, pp. 349-366.

Flanders, H. 1989, *Differential forms with applications to the physical sciences*, New York, Dover

Fleming, W. 1977, *Functions of several variables*, 2nd edition, New York, Springer

Forster, O. 1979, *Analysis 2*, Braunschweig, Vieweg

Forster, O. 2017, *Analysis 3*, 8th edition, Wiesbaden, Springer

Franz, W. 1974, *Topologie II*, 2nd edition, Berlin, de Gruyter, English edition: Algebraic Topology, New York, Ungar, 1968

Fritzsche, K. 2011, *Pfaffsche Formen*, http://www2.math.uni-wuppertal.de/~fritzsch/lectures/ana/an3_k25.pdf

Fues, E. 1927, *Störungsrechnung*, in *Handbuch der Physik. Band V. Grundlagen der Mechanik*, ed. H. Geiger and K. Scheel, Berlin, Springer, pp. 131-177

Gallavotti, G. 1983, *Perturbation theory for classical Hamiltonian systems*, in *Scaling and self-similarity in physics*, Bures-sur-Yvette, 1981/1982, ed. J. Fröhlich, Boston, Birkhäuser and New York, Springer, pp. 359-426

Ganster, M. 2011, *Maß- und Integrationstheorie*, Lecture at TU Graz, https://www.math.tugraz.at/~ganster/lv_masstheorie_ss_11/02_einfache_eigenschaften_von_massen.pdf

Giaquinta, M. and S. Hildebrandt 2004, *Calculus of variations II*, corrected 2nd printing, Berlin, Springer

Godbillon, C. 1983, *Dynamical systems on surfaces*, Berlin, Springer

Goldstein, H., C. Poole and J. Safko 2002, *Classical mechanics*, 3rd edition, San Francisco, Addison-Wesley

Gonchar, A. A. 1995, *Analytic function*, in *Encyclopaedia of Mathematics*, vol. 1, ed. M. Hazewinkel, Dordrecht, Springer, pp. 168-173

Gouvêa, F. Q. 2011, *Was Cantor surprised?*, The American Mathematical Monthly, 118, pp. 198-209

Granas, A. and J. Dugundji 2003, *Fixed point theory*, New York, Springer

Hector, G. and U. Hirsch 1986, *Introduction to the geometry of foliations*, part A, Braunschweig and Wiesbaden, Vieweg

Hilton, P. J. and S. Wylie 1967, *Homology theory*, Cambridge, Cambridge University Press

Hlawka, E. 1990, *Selecta*, Berlin, Springer

Hubbard, J. H. 2007, *The KAM theorem*, in *Kolmogorov's heritage in mathematics*, ed. É. Charpentier, A. Lesne and N. K. Nikolski, Berlin, Springer, pp. 215-238

Jacobi, C. G. J. 1884, *Vorlesungen über Dynamik*, supplement volume to *Gesammelte Werke*, Berlin, G. Reimer

Jech, T. 1997, *Set theory*, Berlin, Springer

Katok, A. and B. Hasselblatt 1995, *Introduction to the modern theory of dynamical systems*, Cambridge, Cambridge University Press

Khesin, B. A. and S. L. Tabachnikov 2012, *Tribute to Vladimir Arnold*, Notices of the American Mathematical Society, 59(3), pp. 378-399, https://doi.org/10.1090/noti810

Khesin, B. A. and S. L. Tabachnikov (eds.) 2014, *Arnold: Swimming against the tide*, Providence, R. I., American Mathematical Society

Kneser, H. 1921, *Untersuchungen zur Quantentheorie*, Mathematische Annalen, 84, pp. 277-302

Kneser, H. 1924, *Reguläre Kurvenscharen auf den Ringflächen*, Mathematische Annalen, 91, pp. 135-154

Kobayashi, S. and K. Nomizu 1963, *Foundations of differential geometry*, New York, Interscience Publishers

Kolmogorov, A. N. 1954, *On conservation of conditionally periodic motions under small perturbations of the Hamiltonian*, Doklady Akademii Nauk SSSR, 98, pp. 527-530

Kolmogorov, A. N. 1957, *General theory of dynamical systems and classical mechanics*, in Proc. of the 1954 intern. congress math., pp. 315-333, Amsterdam, North Holland, for an English translation see the Appendix in Abraham & Marsden (1978)

Krantz, S. G. and H. R. Parks 2002, *A primer of real analytic functions*, 2nd edition, New York, Springer

Krantz, S. G. and H. R. Parks 2013, *The implicit function theorem*, New York, Springer

Kudryavtsev, L. D. 1995, *Function*, in *Encyclopaedia of Mathematics*, vol. 2, ed. M. Hazewinkel, Dordrecht, Springer, pp. 694-699

Kuratowski, K. 1966, *Topology*, vol. 1, New York, Academic Press

Lanczos, C. 1970, *The variational principles of mechanics*, 4th edition, Toronto, University of Toronto Press

Landau, L. D. and E. M. Lifshitz 1976, *Mechanics, Course of theoretical physics*, vol. 1, 3rd edition, Amsterdam, Elsevier Butterworth-Heinemann

Lang, S. 1972, *Differential manifolds*, Reading, Addison-Wesley

Laughlin, G. and J. E. Chambers 2002, *Extrasolar trojans: the viability and detectability of planets in the 1:1 resonance*, The Astronomical Journal, 124, pp. 592-600

Lichtenstein, L. 1931, *Vorlesungen über einige Klassen nichtlinearer Integralgleichungen und Integro-Differentialgleichungen*, Berlin, Springer

Lowenstein, J. H. 2012, *Essentials of Hamiltonian dynamics*, Cambridge, Cambridge University Press

Markushevich, A. I. 1965, *Theory of functions of a complex variable*, vol. 1, Englewood Cliffs, N. J., Prentice-Hall

Milnor, J. 1978, *Analytic proofs of the "hairy ball theorem" and the Brouwer fixed point theorem*, The American Mathematical Monthly, 85, pp. 521-524

Minkowski, H. 1907, *Diophantische Approximationen*, Leipzig, Teubner

Moser, J. K. 1962, *On invariant curves of area-preserving mappings of an annulus*, Nachrichten der Akademie der Wissenschaften in Göttingen, II. Mathematisch-Physikalische Klasse, pp. 1-20

Murray, C. D. 1997, *The Earth's secret companion*, Nature, 387, pp. 651-652

Nordheim, L. and E. Fues 1927, *Die Hamilton-Jacobische Theorie der Dynamik*, in *Handbuch der Physik. Band V. Grundlagen der Mechanik*, ed. H. Geiger and K. Scheel, Berlin, Springer, pp. 91-130

Ostrowski, A. M. 1938, *Sur l'approximation du déterminant de Fredholm par les déterminants des systèmes d'equations linéaires*, Arkiv för Matematik (Stockholm), 26A, pp. 1-15.

Peano, G. 1890, *Sur une courbe, qui remplit toute une aire plane*, Mathematische Annalen, 36, pp. 157-160.

Pöschel, J. 2001, *A lecture on the classical KAM theorem*, Summer School on *Smooth ergodic theory and its applications*, Seattle, July-August 1999, ed. A. Katok, R. de la Llave, Ya. Pesin & H. Weiss, *Proceedings of Symposia in Pure Mathematics*, vol. 69, Providence, R. I., American Mathematical Society, pp. 707-732, available at arxiv.org as arXiv:0908.2234 [this version cited here]

Rothstein, W. & K. Kopfermann 1982, *Funktionentheorie mehrerer komplexer Veränderlicher*, Zürich, Bibliographisches Institut

Rudin, W. 1991, *Functional analysis*, New York, McGraw-Hill

Rüssmann, H. 1979, *Konvergente Reihenentwicklungen in der Störungstheorie der Himmelsmechanik*, in *Selecta Mathematica V*, ed. K. Jacobs, Berlin, Springer, pp. 93-260

Schering, E. 1873, *Hamilton-Jacobische Theorie für Kräfte, deren Maass von der Bewegung der Körper abhangen*, Abhandlungen der königlichen Gesellschaft der Wissenschaften zu Göttingen, 18, pp. 3-54, https://gdz.sub.uni-goettingen.de/id/PPN250442582_0018

Schubert, H. 1964, *Topologie*, Stuttgart, Teubner

Schwartz, J. T. 1969, *Nonlinear functional analysis*, New York, Gordon & Breach

Seifert, H. and W. Threlfall 1934, *Lehrbuch der Topologie*, Leipzig, Teubner

Siegel, C. L. 1942, *Iteration of analytic functions*, Annals of Mathematics (2), 43, pp. 607-612

Siegel, C. L. 1945, *Note on differential equations on the torus*, Annals of Mathematics, 46, pp. 423-428

Siegel, C. L. and J. K. Moser 1971, *Lectures on celestial mechanics*, Berlin, Springer

Sobolev, V. I. 1995, *Mapping*, in *Encyclopaedia of Mathematics*, vol. 3, ed. M. Hazewinkel, Dordrecht, Springer, pp. 745-746

Spivak, M. 1970, *A comprehensive introduction to differential geometry*, vol. 1, Boston, Publish or Perish

Sprindzhuk, V. G. 1995, *Diophantine approximation, Metric theory of –*, in *Encyclopaedia of Mathematics*, vol. 2, ed. M. Hazewinkel, Dordrecht, Springer, pp. 279-280

Sternberg, S. 1964, *Lectures on differential geometry*, Englewood Cliffs, N. J., Prentice-Hall

Sternberg, S. 1969, *Celestial mechanics. Part II*, New York, W. A. Benjamin

Taussky, O. 1949, *A recurring theorem on determinants*, The American Mathematical Monthly, 56, pp. 672-676

Thirring, W. 1977, *Lehrbuch der mathematischen Physik,* vol. 1, *Klassische dynamische Systeme*, Wien, Springer, English translation 1978

Vojta, P. 1993, *Applications of arithmetic algebraic geometry to Diophantine approximations*, in *Arithmetic Algebraic Geometry*, Lectures given at the 2nd Session of the Centro Internazionale Matematico Estivo (C.I.M.E.) held in Trento, Italy, June 24-July 2, 1991, ed. E. Ballico, Berlin, Springer, pp. 164-208

Voronin, S. M. 1995, *Diophantine equations*, in *Encyclopaedia of Mathematics*, vol. 2, ed. M. Hazewinkel, Dordrecht, Springer, pp. 285-287

Wayne, C. E. 1996, *An introduction to KAM theory*, in *Dynamical systems and probabilistic methods in partial differential equations*, ed. P. Deift, C. D. Levermore & C. E. Wayne, *Lectures in Applied Mathematics*, vol. 31, Providence, R.I., American Mathematical Society, pp. 3-29, http://math.bu.edu/people/cew/preprints.html

Whitney, H. 1934, *Analytic extensions of differentiable functions defined in closed sets*, Transactions of the American Mathematical Society, 36, pp. 63-89

Whittaker, E. T. 1917, *A treatise on the analytical dynamics of particles and rigid bodies*, 2nd edition, Cambridge, Cambridge University Press, https://archive.org

Zehnder, E. 1975, *Generalized implicit function theorems with applications to some small divisor problems, I*, Communications on Pure and Applied Mathematics, 28, pp. 91-140

Zehnder, E. 1976, *Generalized implicit function theorems with applications to some small divisor problems, II*, Communications on Pure and Applied Mathematics, 29, pp. 49-111

Zeidler, E. 1976, *Vorlesungen über nichtlineare Funktionalanalysis I. Fixpunktsätze*, Leipzig, Teubner

Zeidler, E. 1986, *Nonlinear functional analysis and its applications I. Fixed-point theorems*, Berlin, Springer

Person Index

Abraham, R., 11, 25, 31, 166
Ahlfors, L. V., 113
Aleksandrov, P. S., 130
Anosov, D. V., 10, 33
Arnold, V. I., ix–xi, 3, 4, 11, 19, 23, 25, 26, 28, 29, 39, 40, 42–44, 47–49, 51–61, 63–67, 71–73, 75–85, 87, 88, 90, 92, 94, 97, 105, 108, 110, 111, 115, 120, 127, 128, 131, 132, 136, 141, 147, 148, 150, 154–156, 159, 161, 171, 175, 178, 180, 182, 184
Arrowsmith, D. K., x
Avez, A., x, 47, 53, 54, 60, 61, 66

Banach, S., 51, 128
Bogoliubov, N. N., 66
Bolzano, B., 172
Borsuk, K., 145
Bott, R., 13, 14
Brent, R. P., 150
Broer, H. W., ix, 43, 48, 127
Brouwer, L. E. J., 115, 129, 134, 136, 138–143, 156, 159

Cantor, G., 43, 65, 69, 70, 107, 142
Cartan, E. J., 11, 13
Cauchy, A.-L., 14, 43, 58, 71, 72, 79, 87, 88, 116, 118, 120, 122, 123, 128, 129, 149, 151, 152, 162
Chambers, J. E., 37

Chierchia, L., ix, x, 47, 64–66, 71, 75, 76, 82
Coriolis, G. G. de, 36
Ćuk, M., 37

de la Llave, R., 6, 47, 49, 52, 59–61, 63, 64, 67, 70, 73, 75, 123, 128
de Morgan, A., 173
Deift, P. A., 11
Descartes, R., 1, 2, 11, 30, 48, 71, 137, 175, 176
Dieudonné, J., 27, 41, 104, 141, 142
Dolzhenko, E. P., 68
Dugundji, J., 51, 129, 131, 136, 145
Dumas, H. S., x, xi, 47, 62, 67

Eliasson, L. H., 52
Euclid, 25, 31, 48, 130, 177
Euler, L., 31, 32, 95

Faltings, G., 60
Fermat, P. de, 60
Fermi, E., x
Feynman, R., 48
Fleming, W., 104
Forster, O., 13, 14, 104, 128
Fourier, J., 40, 47, 49, 50, 52, 57, 60–62, 65–67, 71, 72, 82–84, 115, 118, 119, 121, 175
Franz, W., 136–138
Freudenthal, H., 136
Fritzsche, K., 14
Fues, E., 1, 2, 13, 14, 24, 26, 31

Galilei, G., 6
Gallavotti, G., x
Ganster, M., 172

Gauss, C. F., 139
Giaquinta, M., 13, 27, 28
Godbillon, C., 31
Gödel, K., xi, 150
Goldstein, H., 6
Gonchar, A. A., 41
Gouvêa, F. Q., 142
Granas, A., 51, 129, 131, 136,
        145

Hadamard, J., 106
Hamilton, W. R., ix, xi, 1,
        2, 4, 5, 7, 9–11, 16,
        18, 19, 21, 23–26,
        29, 30, 33, 39, 42,
        48, 49, 51–53, 55–
        58, 67, 70–72, 75,
        76, 81, 82, 89, 90,
        101–104, 106, 108,
        111, 127, 157, 177
Hasselblatt, B., 25
Hector, G., 31
Heyting, A., 136
Hilbert, D., 1, 150
Hildebrandt, S., 13, 27, 28
Hilton, P. J., 136–138
Hirsch, U., 31
Hlawka, E., 60
Hohmann, W., 36, 37
Hubbard, J., xi

Jacobi, C. G. J., 4, 9, 31, 104,
        106, 154, 164, 165

Katok, A., 25
Kepler, J., 36, 37, 48
Khesin, B. A., xi, 11, 148
Klein, F., 31, 32, 137, 139
Kneser, H., 1, 31, 33
Kobayashi, S., 26–28, 30, 164,
        165, 168
Kolmogorov, A. N., ix–xi, 31,
        47–49, 51, 53, 59,
        60, 78
Kopfermann, K., 116

Krantz, S. G., 71, 104, 106,
        127, 128
Kudryavtsev, L. D., 41
Kuratowski, K., 68

Lagrange, J.-L., ix, 1, 11, 35,
        88, 124, 153, 154,
        161, 162, 164, 170,
        171
Lanczos, C., 13
Landau, L. D., 8
Lang, S., 27, 28
Laughlin, G., 37
Legendre, A.-M., 2
Lichtenstein, L., 52
Lie, S., 26, 27, 166
Lifshitz, E. M., 8
Lindstedt, A., 47–49, 51–53
Liouville, J., 10, 11, 19, 23, 25,
        47

Markushevich, A. I., 71, 113
Marsden, J. E., 11, 25, 31, 166
Milnor, J., 25, 142
Minkowski, H., 60
Mitropolsky, Y. A., 66
Moebius, A. F., 137, 139
Mordell, L., 60
Moser, J. K., ix–xi, 47, 48, 60,
        71, 128
Murray, C. D., 38

Nash, J., 48, 67, 71, 128
Newcomb, S., 53, 63
Newton, I., 2, 51, 53, 54, 63,
        64, 72, 127
Noether, E., 150
Nomizu, K., 26–28, 30, 164,
        165, 168
Nordheim, L., 1, 2, 13, 14, 24

Osgood, W. F., 118
Ostrowski, A. M., 150

Parks, H. R., 71, 104, 106, 127,
        128

Pfaff, J. F., 11, 13, 14, 16, 17, 19, 21–23
Pikovski, A., 51
Place, C. M., x
Poincaré, H., 13, 14, 53
Poisson, S. D., 5, 11, 18, 20, 30
Pöschel, J., x, 47, 48, 60, 64–66, 71, 98, 103, 127–129
Pythagoras, 44

Ricci-Curbastro, G., 12
Riemann, B., 14, 48
Roche, E., 35
Rothstein, W., 116
Rüssmann, H., x, 10, 48, 53, 60, 61, 70

Schering, E., 13
Schmidt, E., 52
Schubert, H., 138
Schwartz, J. T., 128
Seifert, H., 132
Sevryuk, M. B., ix, xi, 43, 48, 127
Siegel, C. L., x, 31, 48, 52, 60
Smith, H., 69
Sobolev, S. L., 127
Sobolev, V. I. , 41
Spivak, M., 26–28, 165, 166, 168
Sprindzhuk, V. G., 60
Sternberg, S., x, xi, 26, 27, 48, 53, 66, 67, 104, 128
Stokes, G. G., 13, 139

Tabachnikov, S. L., xi, 11, 148
Taussky, O., 150
Taylor, B., 28, 49, 71, 82, 86, 124
Thirring, W., x, 47
Threlfall, W., 132
Tu, L., 13, 14
Turing, A., xi

Valdinoci, E., 47

Vojta, P., 60
Volterra, V., 69
Voronin, S. M., 60

Wayne, C. E., 48, 61–63, 65, 66, 71, 72, 128, 171, 183
Weierstrass, K., 113, 172
Weyl, H., 71
Whitney, H., 47, 70
Whittaker, E. T., 11, 14, 20
Wylie, S., 136–138

Zehnder, E., 128
Zeidler, E., 131

Subject Index

Acceleration, 36
Accretion, 38
Accumulation point, 68, 70, 172
Action-angle variables, ix, 9, 10, 39, 41, 42, 48, 56, 58, 59, 61, 72, 76, 82, 89, 102, 122
Algebraic topology, xi, 127
Alternating product, 13, 14
Analytic function, ix, x, 33, 39, 41, 42, 54, 55, 57, 61–65, 68, 70, 71, 73, 76, 84, 87, 90, 101, 102, 104, 113, 115, 117–120, 122, 123, 128, 148, 151
Angular frequency, ix, 35, 56
Annulus, 71
Antipodal
    map, 144
    points, 133, 134, 145
    preserving map, 144, 145
    simplex, 144, 145
Arnold-Brouwer lemma, 156, 159
Asteroid, 35, 37
Asymptotic, 104
Average, 51, 71
Averaging principle, 49
Azimuth, 36, 71, 130

Bar, 31, 32
Barycentric refinement, 138, 145
Bifurcation, 137
Bijective, *see* One-to-one
Bilinear invariant, 14
Binary system, 35
Bolzano-Weierstrass theorem, 172

Borsuk theorem, 145
Boundary, 13, 14, 31, 41, 43, 70, 118, 123, 128, 129, 135, 137, 139, 141, 143, 146, 153, 154, 156, 157, 160
    distinguished, 118
    layer, 44, 88, 135, 160, 175, 176, 178–180, 183
    of the image, *see* Image of the Boundary
Bounded domain, function or sequence, 11, 41, 63, 64, 84, 101, 104, 105, 153, 172, 175, 177, 182
Branch of a map, 106, 153
Brouwer lemma, 115, 136, 141, 142

Caltech, 150
Cancellation in series, 47, 52, 84
Canonical
    basis, 30
    transformation, x, 2–6, 8–10, 13, 14, 23, 24, 41, 42, 47–49, 51–53, 55, 56, 58, 59, 63, 65, 67, 70–73, 75, 81, 82, 89, 104, 107, 108, 113, 148, 149, 157, 161, 169, 171, 177, 179
    variables, 2–7, 11, 16, 18, 23, 39, 42, 45, 48, 58, 94, 101
Cantor set, 43, 65, 69, 70, 107
Cartesian
    coordinates, 1, 2, 30, 48,

71, 137, 175, 176
product, 11
Cauchy
    estimate, 71, 72, 79, 87,
        88, 122, 129, 149,
        151, 152
    integral formula, 116, 118,
        122, 123
    sequence, 162
    theorem, 14, 43, 58, 71,
        72, 120, 128
Celestial mechanics, x, 26, 34,
    48, 49, 51
Center of mass, 35
Centrifugal potential, 35
Chain rule, 12, 161, 165, 166
Change of variables or coor-
    dinates, 2, 6, 12, 23,
    24, 53, 59
Chaos, 72
Circle map, 48, 62
Closed
    ball, 43, 44, 153, 157
    domain, 41, 43, 119, 131,
        142, 146, 175, 177
    interval, 69
    manifold, 137, 139
    set, 43, 68–70, 104, 117
    simplex, 136, 138
Closure, 68, 137, 142
Co- and contravariant, 12
Combinatorial lemma, 144, 145
Commutative diagram, 164, 165
Commuting flows, 11, 25–28,
    30
Compact domain or manifold,
    25, 40, 41, 76, 90,
    104, 106, 107, 113,
    139, 171, 172
Complex
    differentiable, 39, 71, 117,
        118
    dimension, 118
    partial differentiable, 117,

118
Concave, 175
Conditionally periodic, ix, 25,
    59
Connected domain, 41, 105,
    117, 120, 135, 137
    doubly, 135, 176
    multiply, 176
    simply, 14, 120, 129, 135,
        175–177
Construction, 31, 34, 63, 101,
    149, 151, 155–158
Continuity of the measure, 172
Continuous deformation, 29,
    130, 134, 135
Contractible domain, 14
Contraction principle, 51, 128
Contradiction, 34, 43, 69, 129,
    132, 134, 136, 139,
    150
Convergence, x, 47, 48, 51–54,
    56, 61, 63, 64, 72, 73,
    75, 79, 84, 97, 106–
    108, 113, 117, 161,
    174
    quadratic, 51–53, 63, 64
    super-, 47, 51, 52, 73
Convex, 14, 154, 175, 176, 178
Convolution, 67
Coordinate-free, 2
Coriolis acceleration, 36
Corner, 29, 31–33
Cotangent vector space, 12
Countable union, 69
Countably additive, 172
Counter-clockwise, 36, 138
Cruithne, 37
Cutoff, *see* Fourier series cutoff
        and Ultraviolet cut-
        off

Decomposition, 61, 101
Degeneracy, 33, 59, 102, 104
Degree theory, xi, 129

Degrees of freedom, ix, 1, 11, 19, 25, 31, 33, 39, 55, 62, 67, 101, 132, 182
Dense, ix, x, 10, 33, 64, 65, 67–69, 111
    in itself, 68
Derived set, 68
Deterministic, x
Diagonally dominant matrix, 150
Diffeomorphism, xi, 25, 27, 41, 42, 57, 58, 62, 63, 65, 67, 70, 72, 75–78, 80, 82–85, 89–92, 102, 104–107, 111, 113, 127, 145, 146, 148, 149, 151, 154–156, 158–161, 166, 167, 169
Differential form, 2, 4, 11–14, 16, 17, 19, 21, 23, 25, 29, 30
    closed, 14
    exact, 14, 15, 19, 21
Diophantine
    approximations, 60
    condition, ix, 37, 47, 48, 51, 57, 60–66, 70, 72, 73, 75, 83, 84, 108, 127, 157, 182
Discontinuous, 63, 65, 143
Discrete set, 68
Distance function, 117
Distribution, 60
Divergence, 64, 177
Domain, 41, 76–78, 89–92, 117
    reduction, 43, 44, 62, 63, 72, 73, 77, 80, 83, 85, 87–89, 98, 105, 106, 119, 127, 149, 154, 156, 157, 159, 160, 175, 178–180, 182–184
    restriction, *see* Domain re-duction
    without interior, *see* Set without interior
Drift, 36, 37
Dual space, 30
Dynamical system, 31, 39, 128

Earth, 35, 37
Edge, 31–33, 71
Energy surface, x
Ergodic, x, 71
    hypothesis, x
Error, 52, 63–65, 73
Essential singularity, 39
Euclidean
    plane, 25, 31, 130
    space, 48, 177
Euler product, 95
Evolution, 5, 58

Family of curves or sets, 27, 31, 33, 172
Fast
    convergence, 48, 51, 62–65, 72, 73
    growth of terms, 52, 60
Fermat theorem, 60
Fixed point theorem, 51, 142, 143
Flow
    in phase space, 11, 25, 27–30
    on manifold, 164, 167–170
Foliation, 11
Formal
    correctness, 170
    order, 53
    solution, 50, 61
    transformation, 83
Fourier
    component, 50, 52, 57, 62, 65, 67, 71, 82–84, 115, 118, 119, 175

series, 40, 47, 49, 57, 60–
        62, 65–67, 72, 82,
        84, 119, 121
series cutoff, 47, 62–68,
        72, 82, 172
theorem, 57
Frequency
    diffeomorphism, xi, 42,
        57, 58, 65, 67, 72, 76,
        77, 80, 83, 89–91,
        102, 104–107, 111,
        127, 145, 146, 154–
        156, 158–160
    map, xi, 50, 70, 72, 81–83,
        88, 102
    shift, 55, 57, 58, 75, 89
    variation lemma, 79, 154
Fundamental
    lemma, 48, 75, 76, 81, 83
    theorem, 53, 55, 58, 75,
        76, 83, 89, 91–94,
        97, 116, 157, 159,
        160

Galilei transformation, 6
Gamma function, 62
Gauss theorem, 139
Generating function, 2, 5, 42,
        48, 62, 63, 65, 70, 75,
        81, 82, 85
Genus, 32
Geometric series, x, 84, 121
Gravity, 34–37, 48

Hadamard theorem, 106
Hairy ball theorem, 25, 142
Hamilton
    equations, 1, 2, 7, 9, 10,
        16, 18, 21, 23–25, 49
    function, ix, 2, 4, 10, 23,
        26, 29, 48, 70, 75
    system, ix, 11, 19, 23–25,
        33, 67, 76, 90
Hamiltonian structure, 26
Harmonic oscillators, 10

Harmonics, 65, 67
Hilfssatz (Brouwer), 136
Hohmann transfer, 36, 37
Holomorphic, 41
Homeomorphism, 31, 137, 154
Homotopy, 130, 134, 137, 138,
        142, 143, 145
    invariance, 127, 129
Horseshoe orbit, 35–37
Hyperplane, 136, 141, 177
Hypersurface, 41

Identity map, 41, 63, 72, 128,
        131, 136, 142
Image of the boundary, 154,
        156
Imbedding, 48
Implicit function theorem, 19,
        47, 48, 51, 104, 127,
        128, 154
Incommensurable, ix, 10, 102,
        108
Index, 130–132, 139
Induction, 55, 63, 93, 98, 99,
        128, 180, 181
Inductive
    lemma, 48, 75, 81
    theorem, 58, 63, 75, 89,
        90, 92–94, 98, 101,
        105–107, 109, 112,
        116
Initial conditions, ix, 2, 34, 59,
        71, 108, 111, 163,
        180
Injective, 113, 128, 129, 146
Inner product, see Scalar prod-
        uct
Integrability, integrable, ix, x,
        10, 11, 20, 21, 25,
        48, 51, 55, 56, 58, 59,
        67, 71, 72, 75, 82, 89,
        101–104, 108, 127,
        148
Integral
    curve, 27, 163

equation, 52
of motion, ix, 11, 19, 23–
    26, 33
Interior, 47, 67–70
    point, 65, 67–70, 136, 137,
        147
Invariant, 12–14, 101, 102, 106,
        107, 111, 127, 129
    torus, ix, 59, 63, 64, 67,
        68, 70, 72, 101, 102,
        107, 111, 113
Inverse
    function, 4, 8, 72, 83, 104,
        113, 127–129, 151,
        153
    function theorem, 104, 106,
        127, 128
    image, 129
Involution, ix, 18–20, 25, 26
Isolated point, 67, 68
Isoperimetric problem, 180
Iteration, 48, 49, 51, 53, 62,
        65, 66, 68, 72, 73, 93,
        119, 127
    index, 42, 171
    lemma, 53, 64
    method, 51, 63, 64
    number, 58
    sequence, 72
    step, xi, 40, 42, 54, 72, 92

Jacobi
    identity, 31
    matrix, 154, 164, 165
Jacobian, 104, 106
Jupiter, 37

KAM, ix–xi, 11, 34, 37, 42, 43,
        47–51, 54, 57, 60,
        62–64, 70–72, 75, 96,
        101–107, 111, 113,
        115, 116, 119, 127,
        128, 148, 154, 175
Kepler ellipse, 36, 37
Kernel, 67

Kinetic energy, 10, 36
Klein bottle, 31, 32, 137, 139

Lagrange
    estimate, 161, 162, 164
    formula, 88, 154, 170, 171
    function, 1
    multiplier, 1
    point, 35
    torus, ix, 11
Legendre transformation, 2
Libration, 36
Lie derivative, 26, 27, 166
Limit
    inferior, 172
    point, 68, 69
    superior, 172
Lindstedt series, 47–49, 51–53
Line net, 31, 32
Linearly independent, 10, 11,
        25, 67, 71
Liouville theorem, 10, 11, 19,
        23, 25, 47
Lobe, 35
Low pass filter, 67
Lower bound, 149, 150

Magnetic field, 138
Majorization, 62
Manifold, 11, 14, 21, 25–27,
        29, 30, 41, 47, 48,
        129, 137, 139, 143,
        165–167, 169, 175
    orientable, 139, 141
    pseudo, 137
Map near identity, 4, 131, 136,
        159
Mapping degree, xi, 127, 129–
        135, 138, 142, 143,
        145
Meagre set, 69
Mean value theorem, 66, 129
Measure, ix, x, 8, 34, 42, 45,
        60, 68–70, 72, 73,
        75, 78, 89, 92, 94, 97,

98, 101, 105, 107, 119, 142, 155, 171, 172, 175–178, 180, 182, 184

Mechanics, x, 2, 5, 6, 11, 13, 29, 47, 60

Metric, 14, 103, 117
    space, 172

Moebius strip, 137, 139

Molecule, ix

Monogenic function, 67, 68

Moon, 35, 37

Mordell conjecture, 60

Multi-index, 82, 117

Multiplicity, 129, 130, 140, 141, 178

Nash-Moser theorem, 128

Neighborhood, 31, 33, 41, 43, 44, 59, 64, 65, 67–71, 117, 133, 163, 170, 171, 177

Nested tori, 10, 33, 72

Newton
    method, 51, 53, 54, 63, 64, 72, 127
    second law, 2

Non-linear, 52, 72, 103

Norm, 47, 55, 58, 88
    1-, 40
    2-, 44, 177
    compatible, 39
    function, 40
    matrix, 39, 110, 161
    maximum, 39, 40, 43, 117, 132, 134, 135, 153, 177, 178
    row-sum, 40
    supremum, 40, 44, 66
    vector, 39, 110, 180–182

Nowhere dense, ix, 43, 68–70, 101, 143

Null-sequence, 89

Number theory, 60

Numerical simulation or mathematics, 53, 72

One form, *see* Pfaff form

One-to-one, 26, 33, 41, 101, 104, 106, 129, 137, 142, 148, 153

Open
    ball, 43, 157
    covering, 104, 172
    disk, 117
    domain, 65, 104, 119, 135
    interval, 69, 70
    set, 26, 33, 41, 43, 65, 68, 70, 104, 117
    simplex, 136, 142

Operator, 54
    integration, 51
    linear, 66
    smoothing, 67, 71

Orbit, 11, 34–37

Order, 48, 49, 51, 104
    2^m, 51, 52
    M^2, 82
    N, 82
    m, 51, 52
    eighth, 58
    first, 2, 5, 19, 52, 56
    fourth, 28, 52, 58
    higher, 56, 66, 75, 161
    low, 67
    second, 5, 28, 52, 53, 56, 58, 59, 86

Orientable manifold, *see* Manifold, orientable

Orientation
    induced, 138, 139
    negative, 138, 140, 141
    of curve, 131, 133
    of simplex, 131, 136, 138–141, 143
    positive, 138, 140, 141

Oriented face, 139

Orthogonal, 25, 50, 57, 83

Parameterization, 10, 102
Pendulum, 102
Perfect set, 68–70
Periodic, ix, x, 11, 25, 26, 35–
        37, 39, 40, 45, 48,
        49, 57, 59, 61, 70, 71,
        76, 82, 90, 101, 102,
        113, 118–120, 122
    multiple, 25, 26
Permutation of simplex cor-
        ners, 138
Perturbation, ix, x, 11, 39, 48,
        51–53, 55–59, 64, 67,
        68, 70–73, 75–77, 81,
        82, 89, 97, 101, 103,
        104, 106, 108, 111,
        127, 148, 154, 161,
        177
    series, 48, 79
    theory, 47, 48, 55, 59, 65,
        104
    theory, linear, 72
Pfaff
    form, 2, 4, 11–14, 16, 17,
        19, 21, 23, 25, 29, 30
    system, 16, 17, 21–23
Phase space, ix, x, 1, 2, 6, 8,
        10, 11, 13, 21, 25, 29,
        31, 33–35, 42, 56,
        67, 75, 82, 101, 103,
        104, 106, 107, 111
    conserved measure, 6–8,
        34, 107
    trajectory, *see* Trajectory
Phase velocity, 29
Piecewise smooth, 118, 175
Planet, ix, 10, 37, 48
Poincaré
    invariant, 13
    lemma, 14
Poisson bracket, 5, 11, 18, 20,
        30
Pole, 39
Polydisk, 117, 118

Polydomain, 117, 118
Polyhedron, 32, 136, 137
    formula, 31, 32
Potential energy, 10
Principle of least action, 1–3
Product topology, 11
Proper function, 106
Pullback, 12
Push forward, 164, 166, 169

Quadrature, 11, 19
Quantum field theory, 48, 67
Quasi-periodic, ix, 10, 71, 101–
        103, 107, 108, 111,
        112

Random, x
Re-entry, 33, 34
Real analytic function, 70, 71
Reduction procedure, 11
Regula falsi, 53, 54
Regular family of curves, 31,
        33
Remainder of series, 66, 119
Removal
    of boundary layer, 70, 175,
        178, 179
    of resonance strip, 180
    of subset, 43, 69
Resonance, 35, 37, 40, 51, 57,
        59–65, 67, 68, 70–
        73, 75, 83, 108, 112,
        157, 159, 177, 179,
        180, 182–184
    strip, 60, 61, 73, 77, 157,
        159, 177, 179, 180,
        182–184
Retracting map, 135
Ricci calculus, 12
Riemann manifold, 14, 48
Roche potential, 35
Rotating frame, 35, 36

Satellite, 35–37
Saturn, 37

Scalar product, 40, 48, 49, 81,
    177
Scheme, 94
    iteration, 63, 64
    Newton, 53, 64, 72
    perturbation, 59
Screw (left- and right-handed),
    138
Second derivative, 21, 80, 149
Sequence of numbers, func-
    tions or domains, x,
    4, 5, 42, 51, 54, 55,
    58, 67, 72, 73, 75, 82,
    89–91, 94, 97, 98,
    103, 106, 107, 113,
    127, 148, 155, 158,
    159, 161, 162, 171–
    174, 177, 179, 183
Set without interior, 47, 67, 68,
    70
Simplex, 136–145
    antipodal, *see* Antipodal
    simplex
    carrier, 137, 141
    closed, *see* Closed sim-
    plex
    collapse, 141, 143
    corner, 136–141, 143–145
    edge, 136
    face, 136–141, 143, 145
    induced orientation, *see*
    Orientation, induced
    open, *see* Open simplex
    orientation, *see* Orienta-
    tion of simplex
    refinement, 138, 142, 145
    side, 136–139, 141, 142,
    144
Simplicial
    approximation, 129, 136–
    139, 141, 143, 145
    approximation theorem, 138
    complex, 136–138, 141
    map, 137, 138, 140–145

Skew-symmetric, 14
Slow
    drift, 36
    frequency change, 58
    perturbation decay, 73
    satellite, 36
Small divisor, 47, 48, 51, 53,
    60, 62–65, 75, 83, 84
Smith-Volterra-Cantor set, 69
Sobolev space, 127
Solar system, 34, 37
Space-filling curve, 142
Sphere
    $(n-1)$-, 143, 144
    n-, 131, 132, 139, 144,
    145, 180
    2-, 25, 144
    with handles, 137
Stabilizing effect of non-linearity,
    103
Stable/unstable, x, 34, 35, 37
Star domain, 14, 15, 21
Stochastic, x
Stokes theorem, 13, 139
Subspace, x, 136
Successive
    approximations, 51, 63
    transformations, 51, 53
Sun, 35, 37, 48
Surjective, 128, 129, 131, 146,
    154
Symplectic (see also Mani-
    fold), 11, 14, 25, 26,
    29, 30, 47

Tadpole orbit, 35, 36
Tangent
    bundle, 26, 166
    space, 26, 30, 165
    vector, 25–27, 164, 166–
    169
Tensor, 14
    metric, 14
Ternary expansion, 69
Thermodynamics, x, 6

Three-body problem, 48
Topological
    anomaly, 142, 143
    equivalence, 137
    invariant, 127
    manifold, 137, 139
    product, 11, 117
    space, 68, 154
Topology, 25, 67–69, 103, 137,
    154
    induced, 117
Torus, ix, x, 10, 11, 14, 25,
    31–33, 42, 56, 59,
    60, 63, 64, 67, 68,
    70–73, 75, 82, 101–
    104, 106–108, 111–
    113, 118, 127, 137,
    139
    deformed, ix, 75, 102–
    104, 111, 113
Trajectory, ix, x, 1, 5, 10, 11,
    25, 31, 33, 34, 51,
    59, 64, 71, 101–104,
    106, 111, 113
    closed, 10, 33, 34, 64
Translation, 6
Triangle
    inequality, 58, 83
    oriented, 138, 139, 144
Triangulation, 131, 137, 141,
    143–145
    basic, 144
    symmetric, 144, 145
Trigonometric polynomial, 65,
    66, 82, 84
Truncation, *see* Fourier series
    cutoff
Type of domain, 42, 90, 182

Ultraviolet cutoff, 67
Uncountable, 69
Uniform convergence, 4, 40,
    48, 60, 107, 113, 161,
    162, 171
Unspooled coordinates, 71

Upper bound, 42, 56, 72, 78,
    86, 147, 149, 178,
    182

Variation, 1–3
    of frequency, 59, 154
Vector field, 25–30, 108, 109,
    163, 164, 166–169
    generates flow, 27, 28, 30,
    164, 167–170
Vector in tangent space, 26, 27,
    164, 166, 169
Vienna circle, 150

Weierstrass theorem, 113
Whitney differentiability, 47,
    70
Wikipedia, 37
Winding number, 130–132
Wolfram
    Alpha, 95
    MathWorld, 70

Zero-velocity curve, 36

Printed in the United States
by Baker & Taylor Publisher Services